The Chestnut Handbook

The Chestnut Handbook

Crop and Forest Management

Edited by
Gabriele Beccaro
Alberto Alma
Giancarlo Bounous
José Gomes-Laranjo

CRC Press
Taylor & Francis Group
Boca Raton London New York

CRC Press is an imprint of the
Taylor & Francis Group, an **informa** business

CRC Press
Taylor & Francis Group
6000 Broken Sound Parkway NW, Suite 300
Boca Raton, FL 33487-2742

First issued in paperback 2021

ISBN-13: 978-1-138-33402-1 (hbk)
ISBN-13: 978-1-03-208430-5 (pbk)

Library of Congress Cataloging-in-Publication Data

Names: Beccaro, G. L., editor.
Title: The chestnut handbook : crop and forest management / [edited by] Gabriele Beccaro, Alberto Alma, Giancarlo Bounous, José Gomes-Laranjo.
Description: Boca Raton : CRC Press, 2020. | Includes bibliographical references. | Summary: "This book presents techniques in chestnut production. It covers planning and management of chestnut from nursery to orchard production, pest control, and interactions with the environment. It is a valuable resource for agronomists, farmers, researchers, and students of agricultural sciences"-- Provided by publisher.
Identifiers: LCCN 2019025188 (print) | LCCN 2019025189 (ebook) | ISBN 9781138334021 (hardback) | ISBN 9780429445606 (adobe pdf)
Subjects: LCSH: Chestnut. | Chestnut--Breeding. | Chestnut--Diseases and pests.
Classification: LCC SB401.C4 C435 2020 (print) | LCC SB401.C4 (ebook) | DDC 634/.53--dc23
LC record available at https://lccn.loc.gov/2019025188
LC ebook record available at https://lccn.loc.gov/2019025189

Visit the Taylor & Francis Web site at
http://www.taylorandfrancis.com

and the CRC Press Web site at
http://www.crcpress.com

Contents

Preface

When we started to write *The Chestnut Handbook*, we contacted many colleagues and friends across the globe, asking them to collaborate to the project. Researchers and professionals with whom we have shared, over many years, the work and passion for the chestnut have enthusiastically contributed, each one with particular experience, expertise, and personal history with the chestnut tree.

The work of almost 40 authors from China, Japan, the United States, Chile, Australia, and Europe has been combined and shaped into 13 chapters of a technical book. Our objective was to make all these authors speak as a single narrator from the beginning to the end of the book.

In this particular moment there is a great lack and a great need for technical information on management of old and new chestnut plantations because the chestnut culture is greatly increasing across the world. The many researchers involved in this editorial project were asked to express viable technical and practical information rather than academic content, with the ambitious goal that the *The Chestnut Handbook* can be a reference volume and a valuable, accessible resource for agronomists, forest experts, farmers, students, researchers, and stakeholders.

Many thanks to all the authors who worked with great enthusiasm on this handbook and to Marta De Biaggi, Giulia Tessa, Isidoro Riondato, and Serena Perrone for additional book editing and reviewing. A very special gratitude goes to Pedro Halçartegaray and Beatriz Cuenca for working so hard sharing their specific technical knowledge on chestnut; to Michele Warmund for her precious contribution and for reviewing many chapters; to Feng Zou and Sogo Nisho for letting us know everything about Chinese and Japanese chestnut; to Valeria Facello for her amazing handmade drawings. Last, but not least, special gratitude goes to Giancarlo Bounous: many contents and collaborations in this book come from his great work and experience on chestnut.

Enjoy it.

Gabriele Beccaro

Editors

Gabriele Beccaro is associate professor of General Arboriculture at the Department of Agricultural, Forestry and Food Sciences of the University of Torino, Italy. He has deepened his expertise and research activities on chestnut, with particular regard for agronomy and propagation topics, in many scientific institutions in the United States, Latin America, China, and Europe. He is the coordinator of the Chestnut R&D Center (Italy), the editor in charge of the magazine *Castanea* (ISSN 2284-4813), and the coordinator of the Chestnut Working Group of the Italian Society for Horticultural Science.

Alberto Alma is full professor of Entomology at the Department of Agricultural, Forestry and Food Sciences, University of Torino, Italy. Specific research activities concerning chestnut protection include: bio-ethology of major and minor chestnut pests; set up of sustainable strategies for chestnut pest control, with special regard to approaches based on sex pheromones; design and realization of biological control of the Asian chestnut gall wasp; set up of parasitoid mass rearing; field multiplication areas, storage, transport, and release of *T. sinensis* in the field; native biocoenosis of the Asian chestnut gall wasp parasitoids; and ethology and life cycle of *T. sinensis*, adaptation to native oak cynipids.

Giancarlo Bounous was appointed full professor of General Arboriculture in 1996 in the former Department of Arboriculture, now Department of Agricultural, Forestry and Food Sciences, University of Torino, Italy, where he served as head of the department. His international profile includes tenure as past chairman of the Chestnut Working Group of the International Society for Horticultural Sciences. He is currently Liaison Leader, Subnetwork on Chestnut, within the framework of the FAO/CIHEAM Inter-regional Cooperative Research Network on Nuts. He has been visiting professor in Europe, Asia, Australia, New Zealand, the United States, and Latin America.

José Gomes-Laranjo is associate professor of Plant Physiology at the University of Trás-os-Montes and Alto Douro, Portugal. He is integrated member on Centre for Research and Technology of Agro-Environment and Biological Sciences, doing research on chestnut issues stressing characterization of varieties, breeding and improvement of chestnut resilience against diseases, and drought and climate change. He is the president of the RefCast-Portuguese Association of Chestnut, coordinator of the Eurocastanea network, and chairman of the Chestnut Working Group of the International Society for Horticultural Sciences.

Contributors

Burak Akyüz
Horticulture Department
Ondokuz Mayis University
Samsun, Turkey

Alberto Alma
Department of Agricultural, Forest and
    Food Sciences
University of Torino
Torino, Italy

Gabriele Beccaro
Department of Agricultural, Forest and
    Food Sciences
University of Torino
Torino, Italy

Elvio Bellini
Centro di Studio e Documentazione del
    Castagno
Firenze, Italy

Roberto Botta
Department of Agricultural, Forest and
    Food Sciences
University of Torino
Torino, Italy

Giancarlo Bounous
Department of Agricultural, Forest and
    Food Sciences
University of Torino
Torino, Italy

Michele Bounous
Vivaio Michele Bounous
Torino, Italy

Jane Casey
Chestnuts Australia Inc.
Myrtleford, Victoria, Australia

Marco Conedera
Insubric Ecosystems Research Group
WSL Swiss Federal Institute for Forest
    Snow and Landscape Research
Cadenazzo, Switzerland

Rita Lourenço Costa
Instituto Nacional de Investigação
    Agrária e Veterinária
Lisbon, Portugal

Jean Coulié
Pépinières Coulié
Le Sorpt, Chasteaux, France

Beatriz Cuenca
Empresa Traformaciòn Agraria S.A.
    (TRAGSA)
Maceda, Ourense, Spain

Marta De Biaggi
Department of Agricultural, Forest and
    Food Sciences
University of Torino
Torino, Italy

Stephanos Diamandis
National Agricoltural Research
    Foundation
Forest Research Institute
Vassilika, Greece

Dario Donno
Department of Agricultural, Forest and
    Food Sciences
University of Torino
Torino, Italy

Valeria Facello
Department of Agricultural, Forest and
   Food Sciences
University of Torino
Grugliasco, Torino, Italy

Chiara Ferracini
Department of Agricultural, Forest and
   Food Sciences
University of Torino
Torino, Italy

Dennis W. Fulbright
Michigan State University
East Lansing, Michigan

Paolo Gonthier
Department of Agricultural, Forest and
   Food Sciences
University of Torino
Torino, Italy

José Gomes-Laranjo
Centro de Investigação e de Tecnologias
   Agro-Ambientais e Biológicas
Universityt of Trás-os-Montes e Alto
   Douro
Vila Real, Portugal

Pedro Halçartegaray Riqué
Vivero Austral
San Bernardo, Chile

Maria Chiara Manetti
CREA - Consiglio per la ricerca in
   agricoltura e l'analisi dell'economia
   agraria
Centro di ricerca Foreste e Legno
Arezzo, Italia

Enrico Marcolin
Dipartimento TESAF
University of Padova
Padova, Italia

Daniela Torello Marinoni
Department of Agricultural, Forest and
   Food Sciences
University of Torino
Torino, Italy

Maria Gabriella Mellano
Department of Agricultural, Forest and
   Food Sciences
University of Torino
Torino, Italy

Fabio Mencarelli
Department for Innovation in
   Biological, Agro-food and Forest
   System (DIBAF)
University of Tuscia
Viterbo, Italy

Sogo Nishio
Pear and Chestnut Breeding Unit
Institute of Fruit Tree and Tea Science
   NARO
Tsukuba, Ibaraki, Japan

Santiago Pereira-Lorenzo
Department of Vegetal Production and
   Engineering Projects
University of Santiago de Compostela
Lugo, Spain

Mario Pividori
Dipartimento TESAF
University of Padova
Padova, Italia

Ana Ramos-Cabrer
Department of Vegetal Production
   and Engineering Projects
University of Santiago de Compostela
Lugo, Spain

Isidoro Riondato
Department of Agricultural, Forest and
    Food Sciences
University of Torino
Torino, Italy

Cécile Robin
INRA
University of Bordeaux
Cestas, France

Ümit Serdar
Horticulture Department
Ondokuz Mayis University
Samsun, Turkey

Giulia Tessa
Department of Agricultural, Forest and
    Food Sciences
University of Torino
Torino, Italy

Andrea Vannini
Department for Innovation
    in Biological, Agro-food
    and Forest System (DIBAF)
University of Tuscia
Viterbo, Italy

Michele Warmund
College of Agriculture, Food & Natural
    Resources
University of Missouri
Columbia, Missouri

Huan Xiong
Key Laboratory of Cultivation and
    Protection for Non-Wood Forest
    Trees, Ministry of Education
Central South University of Forestry
    and Technology
Changsha, Hunan, China

Roberto Zanuttini
Department of Agricultural, Forest and
    Food Sciences
University of Torino
Torino, Italy

Li Zhang
Key Laboratory of Cultivation and
    Protection for Non-Wood Forest
    Trees, Ministry of Education
Central South University of Forestry
    and Technology
Changsha, Hunan, China

Feng Zou
Key Laboratory of Cultivation
    and Protection for Non-Wood Forest
    Trees, Ministry of Education
Central South University of Forestry
    and Technology
Changsha, Hunan, China

1 History
Growing and Using the Chestnut in the World from Past to Present

Giancarlo Bounous and Gabriele Beccaro

Geographically distributed in three main areas throughout the Northern Hemisphere (Asia, Europe, North America), the chestnut (*Castanea* spp.) has an invaluable historical and cultural heritage, a glorious past, but also a promising future, continuing to play an important economical and environmental role in many agroforestry systems. For centuries the nuts provided in rural areas a dietary staple and, when dried, a stored food for the whole year; the wood was used as firewood or building timber.

In Asia, mainly in China, the Chinese chestnuts (*C. mollissima*, *C. henryi*, and *C. seguinii*) are found in wild and cultivated stands and the history of Chinese chestnut cultivation could be traced back to many centuries B.C. Japan, the Korean peninsula and the temperate regions of East Asia are the natural range of the Japanese chestnut (*C. crenata*). *The Chronicles of Japan* (about 700 A.D.) state that in the era of the empress Jito Tenno (645–703 A.D.) the cultivation of chestnuts was promoted in order to fight hunger, and from the Nara to Heian Period (710–1180), the utilization of wild and cultivated chestnuts flourished (Figure 1.1).

Asia Minor, and more specifically the Transcaucasia region, is considered to be the centre for the domestication, cultivation, and dissemination of *C. sativa* (European chestnut or sweet chestnut). During the Roman Empire, thanks to the Roman Legions, the sweet chestnut spread beyond the Italian peninsula into the rest of Europe from East (Romania, Hungary) to West (Spain, Portugal), and to Southern England in the north. From the Middle Ages the chestnut replaced the oak in the European forests, and chestnuts become a staple food. Nowadays the chestnut still dominates forests, and plantations are found in 25 countries and cover a surface of more than 2 million hectares. After a period of abandonment until the end of the twentieth century, parts of old chestnut forests have been restored and new plantations established. In Turkey *C. sativa* is distributed over a broad range both in natural forests and in plantations (Figure 1.2).

In North America, *C. dentata*, a forest giant, was a dominant species in the broadleaf forests along the Appalachian range. The chestnut timber industry vanished about a century ago as the canker blight and ink disease decimated the

1

FIGURE 1.1   An extract from Y. Tanaka's volume, cultivation of chestnut, Meibundou, Tokyo, 1933.

FIGURE 1.2   Wood piles of chestnut timber used to extract tannin, early twentieth century, Italy.

American chestnut. Before being destroyed it furnished small nuts, fuel wood, building timber, and wood products. Nowadays in North America few chestnut plantations of economic importance, based mainly on Asiatic chestnut species or hybrids, have been established.

Into the Southern Hemisphere, mainly in Australia, New Zealand, Chile, and Argentina, *C. sativa* was introduced from Europe during the nineteenth century by the first settlers who brought with them seeds of food plants from their own countries. In Australia the gold rush of the 1850s and 1860s led to the first plantations of chestnut trees. Although the first plantations date back to that period, in Chile and Australia a true chestnut industry was established starting in the middle of the twentieth century and is rapidly growing following new technical interpretations of chestnut cultivation.

Today, the world's leading chestnut producer is China, followed by Korea, Turkey, Italy, Japan, Spain, Portugal, France, and Greece, in decreasing order, but many other countries produce valuable nuts and timber. The market potential appears to be prosperous for the plantations established in suitable environments and the demand continues to outstrip the supply.

Chestnuts are a prized food for an increasingly large market and differ from other nuts in their low fat content, which makes them ideally suited for high complex carbohydrate and low fat diets. The nuts are used to prepare food products of a very healthy quality, often appearing in recipes for balanced dietary schemes. The beneficial effects of the nuts could be the driving force to increase consumption.

Nowadays consumers have a great concern for health, and this unique crop has an outstanding potential for different food products such as vegetables, pastry, dessert, and snacks. The largest part of the production is intended to the fresh market for roasting or boiling which covers a timeframe of six months, with an high demand at the beginning of the harvesting season. Roasted chestnuts in the street are a popular autumn and winter sight in cities all over the world. The best chestnuts, after peeling, are candied or sent to confectioneries to prepare marrons glacés, chestnuts in syrup or in alcohol. Whole peeled chestnuts can become water-packed in tins, dry-packed in glass jars, or vacuum packed. The purée is the base for creams and dairy desserts containing chestnut cream. Peeled and frozen nuts, flakes, beer, or liquor are other processed products high in demand. Dried chestnuts and flour are packed and sold throughout the year.

The increase in demand for environmentally friendly products is a driving force having a positive influence on the market. Technological and aesthetic quality, good resistance to meteoric agents and alterations, and durability make the timber of the species fit for high value products. Timber is used in flooring, building, furniture, and joinery. With coppice small diameter logs, panels, turnery, walking sticks, baskets, charcoal and firewood, garden furniture, traditional fences, vineyard stakes, poles climbing frames, and poles for slope consolidation works are produced. The market offers new products such as finger jointed beams and laminated veneer boards. Tannin is still used for skin tanning, and the wood, after tannin extraction, is reused for the realization of high density fiber panels. The tree is also reported to be a good contributor to carbon sequestration and other ecosystem services.

In addition, this species represents an invaluable bridge between culture and historical heritage: from the "bread tree" of the past, which provided food for generations of farmers, to the present resource, which enables us to satisfy the inborn need to recreate our spirit in a landscape where life can be spent on a human scale. Due to these positive traits, the chestnut landscape and ecosystem must be protected and improved for the great benefits it offers in terms of social welfare. It is a landmark of the history of the people who lived in the areas where the chestnut occurs; the chestnut forests are an unique witness of the liaison among cultivation, memory, and heritage. In old plantations the trees are natural monuments, and these ecosystems contain a large biodiversity and a high cultural heritage created by centuries of human activity. These signs are symbols of a civilization where the elements (dry-stone walls, roads, driers, rural buildings, mills), everyday use objects (furniture, handcrafts), traditions, and ethnic gastronomy should be preserved for leisure and recreation and to improve the quality of life.

2 Botany, Anatomy, and Nut Composition

Gabriele Beccaro, Giancarlo Bounous, Marta De Biaggi, Dario Donno, Daniela Torello Marinoni, Feng Zou, and Maria Gabriella Mellano

CONTENTS

2.1 TAXONOMY AND SPECIES DISTRIBUTION

Fagaceae (*Cupuliferae*) includes six genera (*Castanea, Castanopsis, Fagus, Lithocarpus, Nothofagus, Quercus*) and approximately 1000 species. The genus *Castanea* ($x = 12$, $2n = 24$),[1] chestnuts and chinkapins, is supported as a monophyletic clade closely related to the genus *Castanopsis*.[2]

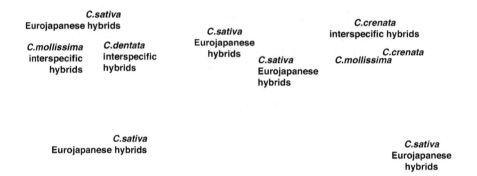

FIGURE 2.1   Main areas of chestnut (*Castanea* spp.) cultivation in the world.

It is widespread in the boreal hemisphere (Figure 2.1) and includes 12 or 13 species depending on the classification (Table 2.1). The natural distribution of the European chestnut (*C. sativa*) covers Europe and all Mediterranean countries. In Asia (China, Korea, Japan, Vietnam) *C. crenata*, *C. mollissima*, *C. seguinii*, and *C. henryi* occur. In North America, *C. dentata* is now rarely present in its native range along the Appalachian Mountains due to the devastating impact of the chestnut blight, and *C. pumila* is found in the southeastern states.

The genus *Castanea* is classified into three sections: *Eucastanon, Balanocastanon*, and *Hypocastanon*, but further revisions are expected due to genetic studies in contrast to the validity of this classification.[3]

Castanea species show high levels of diversity (morphological and ecological traits, vegetative and reproductive habits, nut size, wood characteristics, adaptability and resistance to biotic and abiotic stresses), reflecting the adaptation of the genus to different environmental conditions (Figure 2.2).

TABLE 2.1
Distribution and Use of Chestnut Species

Origin	Section	Species	Common Name	Country	Prevalent Use
Europe	Eucastanon	*C. sativa* Mill.	European or sweet chestnut	Europe, Asia Minor, North Africa	Nut, timber
Asia	Eucastanon	*C. crenata* Seib & Zucc.	Japanese chestnut	Japan, Korea	Nut
		C. mollissima Blume	Chinese chestnut	China	Nut
		C. seguinii Dode		China	Firewood
		C. davidii Dode		China	Firewood
	Hypocastanon	*C. henryi* (Skan) Rehd. & E.H. Wils.	Willow leaf or pearl chestnut	China	Timber

(Continued)

TABLE 2.1 (*Continued*)

Distribution and Use of Chestnut Species

Origin	Section	Species	Common Name	Country	Prevalent Use
America	Eucastanon	*C. dentata* (Marsh.) Borkh.	American chestnut	North America	Timber
	Balanocastanon	*C. pumila* (L.) Mill. var. *pumila*	Allegheny chinkapin	Southeast USA	Nut
		C. pumila (L.) Mill. var. *ozarkensis*	Ozark chinkapin	USA (Arkansas, Missouri, Oklahoma)	Timber
		C. floridana Ashe (Sarg.)	Florida chinkapin	Southeast USA	Ornamental
		C. ashei (Sudw.) Ashe	Ashe chinkapin	Southeast USA	Ornamental
		C. alnifolia Nutt.	Creeping chinkapin	Southern USA (Alabama-Florida)	—
		C. paucispina Ashe		Southern USA (Texas-Louisiana)	—

FIGURE 2.2 Different shapes and growth habits among the different *Castanea* species. (Courtesy of Facello, V.)

Different species are found on very different pedoclimates, but generally prefer deep, soft, acidic soils (pH ranging from 4 to 6.5), temperate climates, and rainfall ranging from 700 to 1500 mm/year. Indeed, chestnut is a dim-light species more adapted to shade and cold north-facing slopes than south-facing ones. At low latitude, chestnut trees may reach 1800 m a.s.l. (Caucasus) and prosper at sea level in Calabria (Italy) or Thessaly (Greece).

Tree shape and form are variable. The genus comprises upright and slender tree species, such as *C. dentata*, as well as smaller species with round foliage and branches that start from the base, or even dwarf shrubs. Plants can live and be regularly productive for centuries, depending on species.

The main species cultivated for fruit are *C. sativa*, *C. mollissima*, and *C. crenata* due to their large nut size, and in particular marrone types (*C. sativa*) are considered the most valuable for nut production. *C. sativa* and *C. dentata* are also used for timber production, whilst many interspecific hybrids are used for nut and timber production or as rootstocks.[4]

2.1.1 EUROPEAN SPECIES

C. sativa (European chestnut or Sweet chestnut) (Figure 2.3) is now the only native species in Mediterranean and Central European regions, being the only species to survive the last glaciation (Würmian). It is a vigorous forest tree that may exceed 400 years of age.

The European chestnut grows in all Mediterranean countries where climatic conditions are suitable, and it is commonly found between 400 and 1000 m above sea level depending on the latitude. Native or cultivated chestnut forests spread from

FIGURE 2.3 European chestnut specimen in Valle di Susa, Italy. (Courtesy of Beccaro, G.)

the Caucasus to Portugal with small patches in Syria and Lebanon, across the inner-Alpine regions reaching Southern England. It is also found in small areas bordering North Africa: Morocco, on Beni-Hoçmar Mountains, Algeria, on Atlas range, Tunisia, probably introduced during French domination. The Canary Islands (27–29°W) represent the southernmost point of the species' distribution range in the northern hemisphere, while its westernmost point is the Azores archipelago (25– 31°W). *C. sativa* was introduced into Chile by European settlers at the beginning of the nineteenth century. The first small extensive plantations in North America occurred in 1773.[5]

Sweet chestnut can suffer from ink disease and canker blight, although some genotypes are partially *Phytophthora* resistant.[6,7] In Europe, the germplasm is very broad and the risk of genetic erosion is high, mostly in marginal or abandoned zones[8,9]; therefore, the preservation of the most interesting genotypes is necessary.

2.1.2 ASIAN SPECIES

C. crenata (Japanese chestnut) spread from its zone of origin, Japan, to Korea and Northeast China, and it was naturalized in South Korea and Taiwan. It can be found between paddy fields and conifer forests, on fertile, recent volcanic soils. It favours mild summer and winter climate, with abundant rainfall (1200–1400 mm/year) in summer. On the southern Japanese Islands, where there is abundant summer rain and mild winters, *C. crenata* may be found at 1300 m a.s.l.

Japanese chestnut is considered one of the most important sources of resistance to *Phytophthora*[10]: in France its germplasm has been used to a large extent in breeding programs to obtain *Phytophthora*-resistant trees.[6] Many cultivars have outstanding nut quality, but they can be attacked by the gall wasp *Dryocosmus kuriphilus*. The Japanese chestnut was introduced into the United States in 1876.[11]

C. mollissima (Chinese chestnut) is the main native nut species in China (Figure 2.4), and owes its name to the thick pubescence on buds and on the abaxial side of leaves. It grows in sub-tropical, temperate-continental, and temperate-maritime regions with hot summers and mild winters. *C. mollissima* was introduced into many countries for its plasticity and adaptability to different pedoclimates. It grows from Jilin Province (41°29'N) close to Korea, down to north of Hainan Island (18°31'N), reaching the central provinces of Shaanxi, Sichuan, Southwards to Yunnan Province, close to the Vietnamese border. It grows from 50 to 2800 m a.s.l. in a wide range of climatic conditions. Several cultivars and local ecotypes have been described, more than 50 of which are cultivated on large surfaces. Chinese chestnut is considered the most resistant *Castanea* species to canker blight and is susceptible to the gall wasp.

Seedlings of *C. mollissima* were introduced by the U.S. Department of Agriculture into the United States and populations are widely grown in the eastern United States. In Connecticut, seedlings regularly produce high quantities of large nuts.

C. seguinii is a small tree or shrub (3–5 m), scattered from 500 to 1000 m a.s.l. in subtropical regions along the Yangzte River, south of the QinLing Mountains, including Shanxi, Henan, Anhui, Zhejiang, Fujian, Jiangxi, Hubei, Hunan, Guizhou, and Yunan Provinces. The very small nuts are harvested for nourishment by rural people and as timber crops.

FIGURE 2.4   A *Castanea mollissima* tree grown in Southern China. (Courtesy of Zou, F.)

C. davidii is considered by some authors a variety of *C. seguinii* based on several affinities.

C. henryi (Figure 2.5) is known as the willow leaved chestnut, or pearl chestnut. The species is adapted to the Chinese warm temperate subtropical climate and it is valuable for fruit and timber production. It grows along Yangtze River Valley and southern regions. The species is considered canker blight resistant, and there is evidence of high genetic variability based on seedling studies.[5]

FIGURE 2.5   *C. henryi* orchard flowering in China. (Courtesy of Zou, F.)

2.1.3 NORTH AMERICAN SPECIES

C. dentata (American chestnut) is native to the eastern United States and Canada (Figure 2.6). Its natural range once extended over more than 200 million acres from southern Ontario and Maine (on the Appalachian Range) to Mississippi. Canker blight led to the complete removal of the species from the forest canopy, while west of the native range it is possible to find adult trees that have escaped the blight.[12] *C. dentata* is the most cold resistant species of the genus.

C. pumila is a polymorphic species, divided into two botanical varieties: *C. pumila* var. *pumila* (Allegheny chinkapin) and *C. pumila* var. *ozarkensis* (Ozark chinkapin).[13,14] It is native to the United States from the east and southeast, occupying small niches on the Ozark mountains of Arkansas, in Missouri, and Oklahoma.[15] Chinkapin tree shapes can be bushy (var. *pumila*), creeping (with some reported to be stoloniferous), or 20 m tall (var. *ozarkensis*).[5]

C. floridana (Florida chinkapin) is a decorative bushy plant native to the southeastern United States from Florida to Texas.

C. ashei (Ashe chinkapin) is spread in North Carolina, Georgia, and Florida.

C. alnifolia is a shrub or creeping chinkapin, native to southern United States, from Alabama to Florida.

C. paucispina is a creeping shrub that grows in Texas and Louisiana.

FIGURE 2.6   An American chestnut seedling. (Courtesy of Beccaro, G.)

2.2 MORPHOLOGY

The following paragraphs summarize the main morphological characteristics of the different *Castanea* species. A list of the main descriptors is reported in Technical Sheet 2.1.

TECHNICAL SHEET 2.1   MAIN MORPHOLOGICAL CHARACTERISTICS

List of the main characteristics and states of expression used to describe species and cultivars. Modified from UPOV Test Guidelines TG/124/4(proj.4).

Tree: growth habit

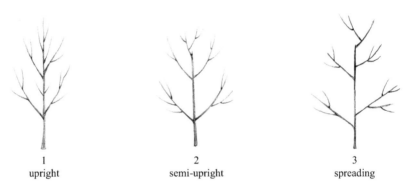

1	2	3
upright	semi-upright	spreading

Current season's shoot: arrangement of leaves

Leaf: shape

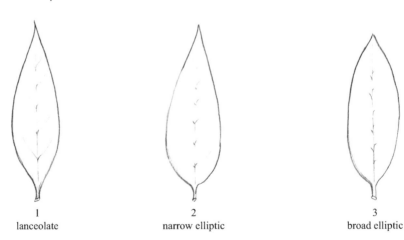

1	2	3
lanceolate	narrow elliptic	broad elliptic

Leaf: shape of base

| 1 | 2 | 3 |
| acute | obtuse | cordate |

Leaf: shape of margin

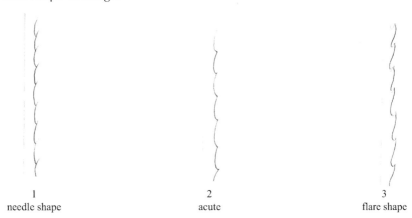

| 1 | 2 | 3 |
| needle shape | acute | flare shape |

Nut: embryonic

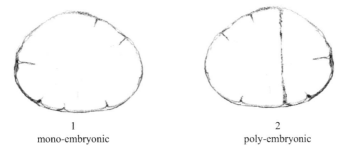

| 1 | 2 |
| mono-embryonic | poly-embryonic |

Nut: degree of penetration of seed coat into embryo

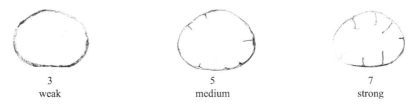

| 3 | 5 | 7 |
| weak | medium | strong |

Nut: shape

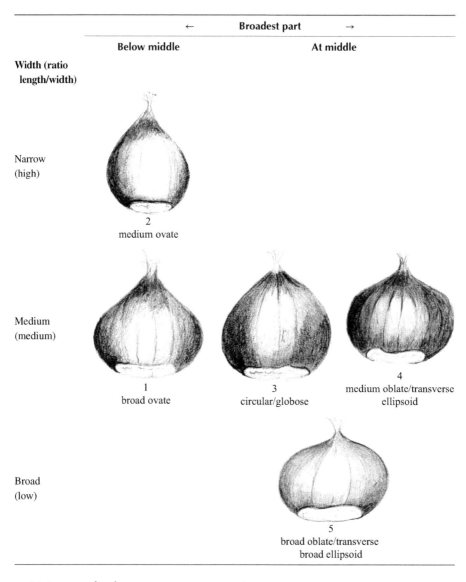

	←	**Broadest part**	→

Width (ratio length/width)

Below middle   **At middle**

Narrow (high)

2
medium ovate

Medium (medium)

1
broad ovate

3
circular/globose

4
medium oblate/transverse ellipsoid

Broad (low)

5
broad oblate/transverse broad ellipsoid

Nut: area of pubescence on upper part

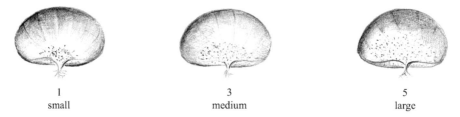

1
small

3
medium

5
large

Nut: area of hilum

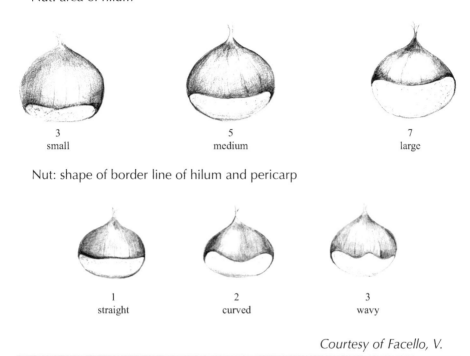

3	5	7
small	medium	large

Nut: shape of border line of hilum and pericarp

1	2	3
straight	curved	wavy

Courtesy of Facello, V.

2.2.1 ROOT SYSTEMS AND MYCORRHIZA

The root system is strong, expanded, and penetrates the soil deeply. Ectomycorrhiza cover abundantly the smallest roots, forming a highly specialized association between plant and fungus. Several mushroom species of high gastronomic interest live in symbiosis with chestnut and are important by-products: *Amanita caesarea, Boletus edulis, Cantharellus cibarius, Lactarius laccata*[16] among Basomycetes, and *Tuber* and *Terfethia* among Ascomycetes.

2.2.2 TRUNK AND BRANCHES

C. sativa is a tall, vigorous, rapid growing tree that can exceed 30 m height, as well as *C. dentata* and *C. henryi*. In young trees the bark is smooth and bright reddish-brown, turning to olive-gray with long lenticels during aging. After 20–25 years the bark has deep longitudinal grooves. The wood has thin, yellowish-white sapwood and brown heartwood and is rich in tannin (5%–7%). Vigorous suckers may grow at the trunk base.

Buds are protected by two budscales; they are ovate, hairless, green-reddish, and fit into leaf scars in a spiral phillotaxy.

C. mollissima is a medium-sized tree (12 m tall and trunk diameter up to 75–80 cm), with trunk branches close to the ground. The trunk is pale grey, smooth when young, with whitish stripes.

C. crenata normally does not exceed 8–10 m in height. Young trees have smooth, thin, brown-olive green bark and lenticels that extend crosswise. Adult trees have brown bark with irregular and deep cracks that sometimes peel in thin strips.

Chinkapins are bushy plants that may reach a height of 6–7 m (*C. floridana* and *C. ashei*), whilst some creeping shrub species do not exceed 30–60 cm (*C. alnifolia* and *C. paucispina*).

2.2.3 LEAVES

Leaves are deciduous, simple, alternate, in a spiral phyllotaxy, with petioles 15–25 mm long (Figure 2.7). The shape is generally elliptical-lanceolate with a round-wedged base, crenate margins, and a narrow acuminate-acute apex. Leaves are in general coriaceous, with a shiny, hairless, and deep green adaxial side, and a dull, light green abaxial side.

European chestnut has large leaves (12–20 × 3–7 cm) with serrated margins. Chinese chestnut leaves (15–20 × 5–7 cm) present a cordate base, irregular and large serrated margins, not well pronounced. The leaves of Japanese chestnut are smaller (9–15 × 3–3.5 cm) with serrated margins showing large and irregular teeth, and short and thick petioles. *C. dentata* leaves are long, narrow, and bright, with big, sharp, and often hooked teeth; they are nearly hairless and paler green than *C. sativa* leaves. The foliage of *C. pumila* is thick with a high variability of leaf forms, size, and colour on the same plant: leaves (4–22 cm long) are sharp, with rough dentate margin, with colours varying from bright yellowish green to light green.

FIGURE 2.7 Chestnut leaves. From left to right: *C. dentata, C. mollissima, C. pumila, C. crenata* 'Tzukuba', *C. sativa* 'Madonna', *C. sativa* marrone type 'Castel del Rio'. (Courtesy of Gamba, G.)

2.2.4 FLOWERS

Chestnut is a monoecious species and on the same tree there are staminate (male) and pistillate (female) flowers in two different kinds of catkins: unisexual, at the base of the shoots, and bisexual on the distal part of the shoots. The number of catkins per shoot varies from 6 to 16.

Staminate flowers occur in a spiral, along the unisexual or bisexual catkins. The catkins are normally 15–20 cm long with a peculiar sweet and musky scent. Each unisexual catkin is composed of flowers gathered in approximately 40 glomerules (axillary cymes) of 3–7 flowers each. The length of the catkins differs according to the species and cultivar (10–40 cm). Each male flower (Figure 2.8a) is formed by a perigonium divided in 8–12 stamens of variable length which carry well-formed anthers, each containing a large quantity of pollen (>1500 grains/anther). Four types of catkins can be distinguished by the length of the stamen filaments: longistaminates (5–7 mm), mesostaminates (3–5 mm), brachistaminates (1–3 mm), and astaminates (without anthers). Almost all catkins of *C. mollissima* and *C. crenata* have longistaminate flowers (Figure 2.9) as do some cultivars of *C. sativa* and many Eurojapanese hybrids. Italian marrone type cultivars generally present astaminate catkins. Pollen has an elongated shape, with a dullish yellow colour. Light-coloured catkins generally have a higher quantity of live pollen than darker-coloured ones. Bisexual catkins often produce non-functional pollen.

(a)

FIGURE 2.8 (a) Male flower with stamens.   (*Continued*)

(b)

FIGURE 2.8 (Continued)   (b) Pistillate flower. (Courtesy of Facello, V.)

FIGURE 2.9   *C. crenata* 'Hakury' in full bloom with longistaminate flowers. (Courtesy of Beccaro, G.)

Pollination is considered both anemophilous[17,18] and entomophilous.[19,20] Anemophilous dispersion of chestnut tree pollen can take place in a range of 30 km[21] to 100 km.[22] However, within 20–30 m, pollen density is modest depending on wind direction and humidity.[9] The period between pollination and nut ripeness varies from 70 to 120 days, from late summer to autumn. The pollination period influences the number of nuts in the bur.[23]

Pistillate flowers (Figure 2.8b) are gathered in globular inflorescences at the base of bisexual (androgynous) catkins. The number of inflorescence per catkin varies from 4 to 5 but only 2 or 3 of them are fertile. Full pistillate flowering is considered when the styles are completely evident and there is maximum receptivity. Each inflorescence generally contains 3 flowers protected by a green, scaled wrapping that is destined to form the cupule which develops into the chestnut bur. There are many ovules in each flower, normally 6, produce a nut with 1 or more seeds.

The chestnut is mainly self-sterile.[17,24,25] Therefore, cross-pollination is needed for effective fertilization. A high degree of sterility in different European, Japanese, and Eurojapanese hybrid cultivars has been observed by Peano et al.[26] *Castanea* genus shows two different kinds of genetic sterility: morphological and genetic (based on incompatibility alleles). Sterility was explained as a shift to dioecy.[25]

2.2.5 NUT

From a botanical point of view the nut is an achene protected by a spiny shell or cupule: the husk or bur (or burr). The spines differ in density and length according to species and cultivar. Chinese chestnut burs present shorter (1–2 cm) and thicker (1 mm) spines if compared to European and American chestnut burs. The green bur turns yellow-brownish reaching maturity. The burs are often paired or clustered on the branch, and may bear 1 to 7 nuts depending on the species and cultivar (generally 3) (Figure 2.10). When the bur achieves full maturity, it opens in two or four valves releasing the nuts. The bur has a subspherical shape, 6–7 cm in diameter in wild trees (even smaller in species such as *C. dentata*) and 10–15 cm in cultivated ones.

The nuts have a smooth and coriaceous pericarp (or shell), which can be light brown or deep brown in colour with more or less evident and raised stripes, and often hairy on the inside. The base of the nut (hilum or hilus scar) is pale brown and

(a)

FIGURE 2.10   Mature burs in autumn containing (a) three nuts (*C. sativa* cv Gabiana).
(*Continued*)

(b)

FIGURE 2.10 (Continued)   Mature burs in autumn containing (b) four nuts (*C. sativa* ×
C. crenata cv Bouche de Betizac). (Courtesy of Bounous, G.)

varies in size. The hilum in *C. crenata* nuts is wide and often reaches the medium
part of the chestnut.

The chestnut apex is composed of the residue of the perianth and of dried styles
forming the torch, which varies in length according to the species and cultivars.
C. mollissima nuts show a long torch covered by a thick, white-cream pubescence.

The seed is wrapped by a thin light-brown pellicle (episperm) that closely adheres
to the seed itself and may penetrate the kernel. Japanese chestnuts present an adherent
pellicle which is difficult to separate from the kernel, whereas Chinese chestnuts show
thin, easy-to-peel pellicle as do the American chestnut and some *C. sativa* cultivars.

The seed can be formed by one (e.g. marrone type) or two cotyledons. The seed
is rich in starch, compact, and whitish inside and yellowish outside. In *C. mollissima*
the pulp is very sweet, but not as the American chestnut, also richer in proteins than
the Japanese and European species.

Nut shape is due to a variety of features including position and number of nuts
inside the bur. When more nuts are present in one single bur, lateral nuts are hemi-
spherical, the central nuts are flattened on one or two sides, and the aborted empty
nuts are flat.

C. sativa nut size ranges from 10 to 30 g, depending on the cultivar. Among
C. sativa cultivars, marrone type is known for its excellent nut quality and sweet
pulp: the shape of the nuts is oblong, with a reddish coloured pericarp with dense and
often raised stripes, and a small semi rectangular shaped hilium. The nut presents
less than 12% kernel division,[27] making the pellicle easily removable.

The nuts of *C. crenata* differ greatly among cultivars: some are the largest in the
genus and can weigh more than 30 g. *C. mollissima* nuts vary from the northern
regions, where the nuts are small (<15 g) with a good and sweet taste, to subtropical
regions, where the nuts are larger (15–20 g) with a high starch content. Very small
nuts characterize *C. seguinii* (1–4 g) and *C. dentata* (1 g) (Table 2.2).

TABLE 2.2
Characteristics of the Most Important Chestnut Species (in Bold the Most Relevant Ones)

Genetic Resources	Characters		
	Nut	Tree	Resistance (R) Susceptibility (s)
C. sativa	**Large size** Adherent pellicle (some cultivars)	**Strong branches** **Good growth habit** **Wood quality**	*Phytophthora* (s) *Cryphonectria* (s) *Dryocosmus* (s)
C. sativa (marrone)	**Large size** **No pellicle intrusion** **Easy to peel** **Sweet flavour** **Good texture** **Ovoid shape** **Small, rectangular hylar scar** **Light coloured shell** **Dark, close stripes**	Lower yield Male sterility More demanding soil and climate requirements	*Phytophthora* (s) *Cryphonectria* (s) *Dryocosmus* (s)
C. crenata	**Very large size (≥30 g)** Adherent pellicle Not sweet, astringent	**Small size (≤15 m)** **High yield** **Precocious bearing** **Early ripening**	*Phytophthora* **(R)** *Cryphonectria* **(R)** (moderate) *Dryocosmus* (s) (high) Spring frost (s)
C. mollissima	**Weight (10–30 g)** **Sweetness, flavour, protein content** **No pellicle intrusion** **Thin pellicle** **Easily removed pellicle** High variable size	**Medium size (≤20 m)** **Semi-upright habit** **Early ripening (variable)** Precocious (variable) Two crops/year (in sub-tropical areas) (variable) **Good pollinizer**	*Phytophthora* **(R)** *Cryphonectria* **(R)** (variable) *Dryocosmus* (s)
C. dentata	**Very sweet** **Non-astringent** **Easy to peel** Very small (300 nuts/kg)	**Fast, straight growth with strong central leader** **Self pruning** **Well coppiced**	*Cryphonectria* (s) (high) **Frost or cold (−35°C) (R)**
C. seguinii	Small size **Very prolonged blooming and ripening period** **Very precocious**	**Small, medium size** **Precocious flowering** **Ever bearing** 2 crops/year (some clones) **Chain of 10–20 burs** (some clones)	*Cryphonectria* **(R)** *Dryocosmus* (s)

(Continued)

TABLE 2.2 (*Continued*)

Characteristics of the Most Important Chestnut Species (in Bold the Most Relevant Ones)

	Characters		
Genetic Resources	**Nut**	**Tree**	**Resistance (R) Susceptibility (s)**
C. pumila	Very small	**Moderate size**	*Cryphonectria* (**R**)
	Single nut burs	**Stoloniferous clones**	(partial)
	Sweet, flavourful	**Prolific suckering**	**Warmer temperate**
	Very precocious	**ability**	**climates (R)**
		Soft spined burs	**Quickly replacing**
		Suitable for warm	**blighted stems**
		climate	
C. henryi	Single nut burs	**Fast growth**	*Cryphonectria* (**R**)
	Very small	**Straight trunk**	
		Good wood	
		Suitable for warm	
		temperate or tropical	
		climates	

Source: Bounous, G. and Marinoni, D.T., *Hortic. Rev.*, 31, 291–347, 2005.

2.3   PHENOLOGICAL PHASES

Like other fruit species, the chestnut presents an intense vegetative and reproductive activity from early spring to autumn, depending on latitude and altitude. Bloom, nut maturation, and harvest time differ among the species and cultivars of the *Castanea* genus. The chestnut is, in general, late blooming compared to other temperate climate tree species: flowers typically reach anthesis in mid-summer, with differences among species and according to geographic location. Across the northern hemisphere, European chestnuts bloom typically from mid-May throughout June (early May for Japanese chestnut), American chestnut from June to July, whilst *C. seguinii* has early flowering and continues to flower throughout the bearing season until frost. *C. mollissima* varies bloom period from northern China (from early-June through the end of June) to southern China (May). In the southern hemisphere chestnuts normally bloom from November to December.

As a consequence, *C. mollissima* and *C. crenata* nuts, together with Eurojapanese hybrids, mature earlier than other species, while marrone type (*C. sativa*) nuts mature later in the autumn.

1. Bud swelling

   Along a branch both vegetative and mixed buds occur. Bud swelling begins as temperatures rise in spring.

Courtesy of Beccaro G.

2. Bud break

   During this phase the swollen buds break the perula and green leaf tips appear.

Courtesy of Beccaro G.

3. First leaves begin to unfold.

Courtesy of Gamba, G.

4. Catkins appearance. Chestnuts are generally protandrous (pistillate flowers open after staminate). Flowers bloom only after leaves are completely open. Unisexual catkins blossom before bisexual ones, a phenomenon called duodicogamy.

Courtesy of Beccaro G.

5. Visible glomerules along the catkin with catkins reaching full length.

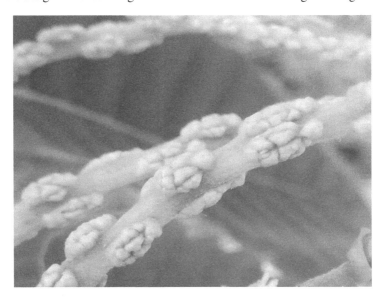

Courtesy of Beccaro G.

6. Stamen appearance.

Courtesy of Gamba, G.

7. Pistillate flower appearance and growth. As each flower becomes receptive, the cluster of styles becomes visible.

Courtesy of Beccaro G.

8. Pollen emission and full anthesis.

Courtesy of Beccaro G.

9. Catkin senescence and bur development.
   Pollen production stops roughly 3–4 weeks after the start of bloom, while unisexual catkins drop, female flower stigmas cease receptivity, and the fertilized flowers grow into green spiny burs containing the ovaries developing into chestnuts.

Courtesy of Beccaro G.

10. Nut ripening and browning of chestnut burs. Burs split letting the nuts out. (a) In some species and cultivars burs split and let nuts out while still attached to the branches (e.g. *C. mollissima*); (b) in other cases ripening burs open when they fall on the ground.

(a)

Courtesy of Beccaro G.

(b)

2.4   NUT COMPOSITION

2.4.1   CHEMICAL COMPOSITION

2.4.1.1   Carbohydrates

Carbohydrates, mainly starch and sucrose, are the main components of the chestnut (30–40 g/100 g of the edible fresh fruit). Chestnuts present higher sucrose content (5–10 g/100 g of fresh weight) than wheat, walnuts, and potatoes. The high starch content (25–30 g/100 g of fresh weight) should be taken into account when selecting fruit for the food industry.[28]

2.4.1.2   Fibre

Fibre content in chestnuts ranges from 7 to 8 g per 100 g of fresh nuts, and it is responsible for nut structure determining its consistency, important in assessing the product quality.[29]

2.4.1.3   Protein

The protein content (1.5–4.5 g/100 g of fresh weight) is similar to milk, while pro-lamin and glutenin (pro-gluten molecules) are absent. For this reason, chestnut flour can be used for bread production only if mixed together with cereal or rye flour. The amino acid content is (g amino acid/100 g chestnut): alanine 0.14–0.25, arginine 0.14–0.44, aspartic acid 0.33–0.85, cystine 0.06–0.11, glutamic acid 0.25–0.62, gly-cine 0.11–0.23, histidine 0.05–0.13, isoleucine 0.08–0.19, leucine 0.11–0.30, lysine 0.12–0.27, methionine 0.05–0.10, phenylalanine 0.08–0.22, proline 0.10–0.21, serine 0.10–0.21, threonine 0.07–0.19, tryptophan 0.02–0.05, tyrosine 0.06–0.16, and valine 0.11–0.26.[30] Globulins are the main proteins, with a high albumin content. Chestnut proteins are high in lysine and threonine, and methionine is scarce. Chestnuts also contain considerable amounts of γ-amino butyric acid (GABA).

2.4.1.4 Lipids

Unlike other dry fruits (such as walnuts, hazelnuts, and almonds) rich in fatty acids and lipids, chestnuts show low fat values (0.5–2.0 g/100 g) and are free of cholesterol.[31] The essential fatty acid content (linoleic and linolenic acids) is similar to potatoes and wheat, accounting for about 65% of the total fat content.[32]

The European chestnut has a high saturated fatty acid (SFA) and polyunsaturated fatty acid (PUFA) content (about 20% and 45%, respectively). On the other hand, Japanese and Chinese chestnuts present a higher monounsaturated fatty acid (MUFA) content (about 55%–60%) than the European chestnut (about 40%).[33] Fatty acids in the chestnut are the palmitic (16:0), stearic (18:0), oleic (18:1ω9), linoleic (18:2ω6), and α-linolenic (18:3ω3) acids. *C. mollissima* and *C. dentata* chestnuts have the highest unsaturated fatty acids content (89% and 87%, respectively), while palmitic acid is the predominant SFA. Stearic acid shows relevant quantities only in American and European chestnuts, while small amounts are present in *C. mollissima*. Oleic acid is the main fatty acid in American and Chinese chestnuts, while linoleic acid is the main fatty acid in the European chestnut.

2.4.1.5 Organic Acids

The main organic acids present in chestnuts are the oxalic (1–30 mg/100 g of fresh weight), citric (100–400 mg/100 g), malic (10–150 mg/100 g), quinic (50– 200 mg/100 g), and fumaric (0.05–2 mg/100 g) acids.[34]

2.4.1.6 Mineral Elements

American chestnuts show an higher potassium content (500 mg/100 g fresh weight) than the other major species.[30] Magnesium, phosphorus, and manganese contents in the Chinese chestnut (about 84, 96, and 2 mg/100 g, respectively) are higher than in European (about 30, 40, and 0.5 mg/100 g, respectively) and Japanese ones (50, 70, and 1.5 mg/100 g, respectively). The Japanese chestnut contains a high amount of calcium, iron, sodium, zinc, and copper contents (about 30, 1.5, 15, 1, and 0.5 mg/100 g, respectively). Sodium content in American, Chinese, and European species is about 3 mg/100 g, while calcium content in *C. dentata* nuts varies between 20 and 25 mg/100 g.[30]

2.4.1.7 Polyphenolic Compounds

American and Chinese chestnuts contain about 500 μg of phenolic acids/100 g of fresh fruits. Gallic acid is the predominant phenolic acid present in these species (about 400 μg/100 g). Protocatechuic, caffeic, p-hydroxybenzoic, and syringic acids are present in low concentrations.[35] In the chestnut the highest concentration of the total phenolics is in the episperm (1000–1500 mg/100 g of fresh weight), followed by the pericarp (300–500 mg/100 g), while the lowest phenolics content is in the endosperm (100–200 mg/100 g of fresh weight). Flavonoids are found in small amounts (flavonols 4.5–15.0 mg/100 g of edible portion, catechin 0.1–3.0 mg/100 g, and epicatechin 1.5–8.0 mg/100 g). It has been observed that total phenolics decreased after 6 months of storage at 4°C and −20°C.[36]

The pellicle (skin) of the chestnut is rich in tannins, a group of phenolics, which contributes to its astringency sensation. The tannins in chestnut flesh are mainly composed of gallotannins (3,6-digalloyl glucose, pyrogallol, and resorcinol) and ellagitannins (castalagin and vescalagin). Tannins content varies from 8% to 70% in the cotyledons and from 0.3% to 2% in the pericarp.[37]

2.4.1.8 Vitamins

Chestnuts contain considerable amounts of vitamin C, folate, and vitamin A. Vitamin C content is higher in the European chestnut (about 30 mg/100 g of fresh fruit), compared to the average of Chinese and Japanese varieties (25 and 16 mg/100 g, respectively). Folate and vitamin A contents are on average higher in the Chinese chestnut (70 and 10 μg/100 g, respectively) compared to other species.[38] Chestnuts also contain thiamine (B1), riboflavin (B2), niacin (B3), pantothenic acid (B5), and pyridoxine (B6). Group B vitamins are thermo-stable (are not destroyed by cooking) and form an active part of different coenzymes that participate in essential metabolic functions. Vitamin B1, although only present in modest quantities (0.1 mg/100 g of fresh fruit) maintains muscle tone during physical activity. Chestnuts present a trace of vitamin E and B group vitamins.[39] European chestnuts contain about 0.5 mg/100 g of δ-tocopherol and 8.0 mg/100 g of γ-tocopherol (fresh weight).[40]

2.4.2 ALLERGENS

There is limited information available on the allergenicity of chestnuts. Some studies report that the chestnut is the third most prevalent food allergen in both adult and paediatric allergy patients. Boiling the nuts has been reported to decrease the allergenic potency.[41]

2.4.3 NUTRITIONAL VALUE AND HEALTH PROPERTIES

Thanks to its nutritional and sensorial traits and potential health effects, the chestnut plays an important role in the human diet and can be considered a health-promoting food (Technical sheet 2.2). Compared to other tree nuts, the chestnut has a low fat content and it is cholesterol-free. It provides a good source of essential fatty acids, and is useful to prevent cardiovascular diseases in adults and brain and retina development in infants. The significant presence of PUFA, as linoleic acid, may help to reduce cholesterol and prevent coronary heart disease.[42] Chestnuts are an excellent energy source and they are ideal for physical and mental stress conditions, but not recommended for diabetics. Moreover, chestnuts could be considered a valid alternative food for people allergic to lactose by cow milk or individuals with cereal intolerance in the preparation of sweet products and soups.[12]

TECHNICAL SHEET 2.2   SENSORY ANALYSIS

Sensory evaluation, now widely applied to a large range of food,[43] can be conducted also on the chestnut in order to identify and quantify the organoleptic traits of the different cultivars.[44]

In order to guarantee a common lexicon and methodology of evaluation, specific training sessions must be carried out before the evaluation. During the training sessions, at least 12 selected panellists work in groups and individually.

After each evaluation, a discussion is held to select the appropriate set of descriptors.[45] Each descriptor (peelability, seed colour, chestnut smell intensity, sweet taste intensity, salty and bitter intensity, flouriness, and chestnut aroma intensity) is selected and defined according to bibliographic references,[46,47] and a common glossary is evaluated on a continuous scale partially structured into 10 segments.[46]

Sensory tests are conducted in a specific sensory laboratory, applying a quantitative descriptive analysis (QDA) as the analytical-descriptive method.[46] This kind of analysis has several benefits: better definition of the criteria for the raw materials selection, fruit quality control, and optimization of the consumption modalities. Besides, together with the chemical analysis, it has a primary role for the description, evaluation, and comparison of the chestnut cultivars.

Lexicon of Chestnut Sensory Attributes and Their Associated Training Methodology and Reference Standards

Attribute	Description	Graduated Scale	Training Methodology
Peelability	The ease of detachment of epicarp and episperm from the fruit flesh	0 = hard; 10 = easy	At least 30 individual and shared tests on different cultivars
Seed colour	Intensity of the seed colour	0 = dark; 10 = clear	At least 30 individual and shared tests on different cultivars
Chestnut smell intensity	Intensity of the sweet chestnut odour	0 = null; 10 = very high	At least 30 individual and shared tests on different cultivars
Sweet taste intensity	Intensity of the sweet savour	0 = null; 10 = very high	Individual and shared tests with water and sucrose solutions at concentrations (0.5–6.0 g/L)
Salty taste intensity	Intensity of the salty savour	0 = null; 10 = very high	Individual and shared tests with water and sodium chloride solutions at concentrations (0.2–2.4 g/L)
Bitter taste intensity	Intensity of the bitter savour	0 = null; 10 = very high	Individual and shared tests with water and sodium caffeine at concentrations (0.034–0.20 g/L)
Flouriness	The amount of dry, fine, powdery particles that coat the mouth during chewing	0 = null; 10 = very high	At least 30 individual and shared tests on different cultivars and specifics test on reference standards (raw carrots for absent/low values and canned beans for extreme/high values)
Chestnut aroma intensity	Intensity of the sweet chestnut aroma	0 = null; 10 = very high	Tasters experience; at least 30 shared tests on different cultivars

Source: Donno, D. et al., J. *Food Quality*, 35, 169–181, 2012; Piggott, J.R. et al., *Int. J. Food Sci. Technol.*, 33, 7–12, 1998; Mellano, M.G. et al., *Castanea*, 2, 2017; Mellano, M. et al., *Italus Hortus*, 12, 64, 2005; Zeppa, G. et al., *Quaderni della Regione Piemonte*, 38, 35–39, 2003.

Nuts are generally recommended along with vegetables, fruits, and cereals to increase the fibre intake of consumers. The chestnut is also a potential health-promoting food because of its vitamins (in particular, C and B group vitamins), mineral elements, aminoacids, and polyphenolic compounds. Arginine, potassium, copper, and magnesium also contribute to its positive nutritional value. Low sodium content is a further chestnut strength, as low sodium diets are recommended in order to reduce blood pressure.[48]

Nutritional value considerably varies depending on cooking and preparation methods. If boiled, chestnut humidity increases and energy value decreases (about −25%). In roasted chestnuts the humidity drops to about 42% (−20%). Cooking also alters starch, sucrose, lipid, and protein contents, and potassium and magnesium contents (not calcium) are reduced by boiling. When chestnuts are dried, protein content increases (5%–6%). Carbohydrate levels also increase (approximately 60 g/100 g of edible matter). Dried chestnuts show modest contents of sodium (15 mg/100 g), iron (1.5–2.0 mg per 100 g), calcium (40–50 mg/100 g), and potassium (about 740 mg/100 g). In chestnut flour fibre content is very high (14%), with the prevalence of insoluble portion (90% of total fibre).[49,50]

REFERENCES

1. Jaynes, R.A. 1962. Chestnut chromosomes. *Forest Science* 8(4): 372–377.
2. Manos, P.S., Zhou, Z.-K. and Cannon, C.H. 2001. Systematics of Fagaceae: Phylogenetic tests of reproductive trait evolution. *International Journal of Plant Sciences* 162(6): 1361–1379.
3. Santamour, F., McArdle, A. and Jaynes, R. 1986. Cambial isoperoxidase patterns in *Castanea*. *Journal of Environmental Horticulture* 4(1): 14–16.
4. Pereira-Lorenzo, S., Ballester, A., Corredoira, E., Vieitez, A.M., Agnanostakis, S., Costa, R., Bounous, G., Botta, R., Beccaro, G.L. and Kubisiak, T.L. 2012. Chestnut. In *Fruit Breeding*, Badenes, M.L. and Byrne, D.H. (Eds.), pp. 729–769. Springer, New York.
5. Bounous, G. 2014. *Il castagno: risorsa multifunzionale in Italia e nel mondo*. Edagricole, Bologna, p. 420.
6. Salesses, G., Chapa, J. and Chazernas, P. 1993a. Screening and breeding for ink disease resistance. *Proceedings of the International Congress of Chestnut*, Spoleto (PG), Italy, p. 5.
7. Salesses, G., Ronco, L., Chauvin, J. and Chapa, J. 1993b. Amélioration génétique du chataignier, Mise au point de tests d'evaluation du comportement vis-à-vis de la maladie de léncre. *Lárboriculture Fruitiere* 458: 23–31.
8. Grassi, G. 1992. Individuazione, valutazione e conservazione di biotipi e cultivar di castagno da frutto. *Atti Convegno "Germoplasma Frutticolo," Alghero (SS), Italy*, pp. 603–606.
9. Pisani, P.L. and Rinaldelli, E. 1991. Alcuni aspetti della biologia fiorale del castagno. *Frutticoltura* 52: 6.
10. Rutter, P.A., Miller, G. and Payne, J.A. 1991. Chestnuts (*Castanea*). *Acta Horticulturae* 290: 28.
11. Rosengarten Jr, F. 2004. *The Book of Edible Nuts*. Dover Publications, New York.
12. Bounous, G. and Marinoni, D.T. 2005. Chestnut: Botany, horticulture, and utilization. *Horticultural Reviews* 31: 291–347.
13. Johnson, G. 1987. Chinquapins: Taxonomy, distribution, ecology, and importance. *Annual Report of the Northern Nut Growers Association*, Ohio, USA.

14. Johnson, G.P. 1988. Revision of Castanea sect Balanocastanon (Fagaceae). *Journal of the Arnold Arboretum* 69: 25–49.
15. Camus, A. 1929. Les Châtaigniers. Monographie des genres Castanea et Castanopsis. *Enciclopedie economique de sylviculture.* Lechevalier, Paris. III, p. 604.
16. Meotto, F., Pellegrino, S. and Bounous, G. 1999. Evolution of *Amanita caesarea* (Scop.: Fr.) Pers. and *Boletus edulis* Bull.: Fr. synthetic ectomycorrhizae on European chestnut (*Castanea sativa* Mill.) seedlings under field conditions. *Acta Horticulturae* 494: 201–206.
17. Clapper, R. 1954. Chestnut breeding, techniques and results: II. Inheritance of characters, breeding for vigor, and mutations. *Journal of Heredity* 45(4): 201–208.
18. Porsch, O. 1950. Geschichtliche lebenswertung der kastanienblüte. *Oesterreichen Bot. Z.* B97: 2.
19. Morettini, A. 1949. *Biologia fiorale del castagno.* Stab. Tipográfico Ramo Editoriale Degli Agricoltori.
20. Breviglieri, N. 1951. Ricerche sulla biologia fiorale e di fruttificazione della Castanea sativa e Castanea crenata nel territorio di Vallombrosa. *Centro di Studio Sul Castagno* 1: 15–49.
21. Peeters, A.G. and Zoller, H. 1988. Long range transport of *Castanea sativa* pollen. *Grana* 27(3): 203–207.
22. Tampieri, F., Mandrioli, P. and Puppi, G. 1977. Medium range transport of airborne pollen. *Agricultural Meteorology* 18(1): 9–20.
23. Shimura, I., Yasuno, M. and Otomo, C. 1971. Studies on the breeding behaviors of several characters in chestnuts, *Castanea* spp.: II. Effects of the pollination time on the number of nuts in the burr. *Japanese Journal of Breeding* 21(2): 77–80.
24. McKay, J.W. 1942. Self-sterility in the Chinese chestnut (*Castanea mollissima*). *Proceedings of the American Society for Horticultural Science* 41: 156–160.
25. Jaynes, R.A. 1975. *Chestnuts.* Purdue University Press, West Lafayette, IN.
26. Peano, C., Bounous, G. and Paglietta, R. 1990. Contributo allo studio della biologia fiorale e di fruttificazione di cultivar europee, orientali ed ibride del genere *Castanea* Mill, 83–99.
27. Bergonoux, F., Verlhac, A., Breisch, H. and Chapa, J. 1978. Le châtaigner, production et culture. *Comité National Interprofessionel de la Chataigne et du Marron, Paris. [In French].*
28. Zhu, F. 2017. Properties and food uses of chestnut flour and starch. *Food and Bioprocess Technology* 10(7): 1173–1191.
29. Gonçalves, B., Borges, O., Costa, H.S., Bennett, R., Santos, M. and Silva, A.P. 2010. Metabolite composition of chestnut (*Castanea sativa* Mill.) upon cooking: Proximate analysis, fibre, organic acids and phenolics. *Food Chemistry* 122(1): 154–160.
30. Korel, F. and Balaban, M.Ö. 2008. Chemical composition and health aspects of chestnut (*Castanea* spp.). In *Tree Nuts: Composition, Phytochemicals, and Health Effects*, Alasalvar, C. and Shahidi, F., (Eds.), p. 171. CRC Press Taylor & Francis Group, Boca Raton, FL.
31. Borges, O.P., Carvalho, J.S., Correia, P.R. and Silva, A.P. 2007. Lipid and fatty acid profiles of *Castanea sativa* Mill. Chestnuts of 17 native Portuguese cultivars. *Journal of Food Composition and Analysis* 20(2): 80–89.
32. Künsch, U., Schärer, H., Patrian, B., Höhn, E., Conedera, M., Sassella, A., Jermini, M. and Jelmini, G. 2001. Effects of roasting on chemical composition and quality of different chestnut (*Castanea sativa* Mill) varieties. *Journal of the Science of Food and Agriculture* 81(11): 1106–1112.
33. Vekiari, S.A., Panagou, E. and Mallidis, C. 2006. Compositional analysis of chestnuts in Mediterranean countries. *Advances in Horticultural Science* 20(1): 90–95.
34. Carocho, M., Barros, L., Antonio, A.L., Barreira, J.C.M., Bento, A., Kaluska, I. and Ferreira, I. 2013. Analysis of organic acids in electron beam irradiated chestnuts (*Castanea sativa* Mill.): Effects of radiation dose and storage time. *Food Chemical Toxicology* 55: 348–352.

35. Senter, S.D., Payne, J.A., Miller, G. and Anagnostakis, S.L. 1994. Comparison of total lipids, fatty-acids, sugars and nonvolatile organic-acids in nuts from 4 *Castanea* species. *Journal of the Science of Food and Agriculture* 65(2): 223–227.

36. De Vasconcelos, M.C., Bennett, R.N., Rosa, E.A. and Ferreira-Cardoso, J.V. 2010. Composition of European chestnut (*Castanea sativa* Mill.) and association with health effects: Fresh and processed products. *Journal of the Science of Food and Agriculture* 90(10): 1578–1589.

37. Hwang, J.-Y., Hwang, I.-K. and Park, J.-B. 2001. Analysis of physicochemical factors related to the automatic pellicle removal in Korean chestnut (*Castanea crenata*). *Journal of Agricultural and Food Chemistry* 49(12): 6045–6049.

38. Neri, L., Dimitri, G. and Sacchetti, G. 2010. Chemical composition and antioxidant activity of cured chestnuts from three sweet chestnut (*Castanea sativa* Mill.) ecotypes from Italy. *Journal of Food Composition and Analysis* 23(1): 23–29.

39. Salvini, S., Parpinel, M., Gnagnarella, P., Maisonneuve, P. and Turrini, A. 1998. Banca dati di composizione degli alimenti per studi epidemiologici in Italia. Istituto europeo di oncologia, stampa, Milano, Italy.

40. Bellini, E., Giordani, E., Marinelli, C. and Perucca, B. 2004. Marrone del Mugello PGI chestnut nutritional and organoleptic quality. *III International Chestnut Congress Acta Horitculturae* 693: 97–102.

41. Lee, S.-K., Yoon, S.-H., Kim, S.-H., Choi, J.-H. and Park, H.-S. 2005. Chestnut as a food allergen: Identification of major allergens. *Journal of Korean Medical Science* 20(4): 573–578.

42. Connor, W.E. 1997. The beneficial effects of omega-3 fatty acids: Cardiovascular disease and neurodevelopment. *Current Opinion in Lipidology* 8(1): 1–3.

43. Donno, D., Beccaro, G., Mellano, M., Torello Marinoni, D., Cerutti, A., Canterino, S. and Bounous, G. 2012. Application of sensory, nutraceutical and genetic techniques to create a quality profile of ancient apple cultivars. *Journal of Food Quality* 35(3): 169–181.

44. Piggott, J.R., Simpson, S.J. and Williams, S.A. 1998. Sensory analysis. *International Journal of Food Science & Technology* 33(1): 7–12.

45. Mellano, M.G., Rapalino, S. and Donno, D. 2017. Sensory profiles of *Castanea sativa* cultivars and Eurojapanese hybrids. *Castanea* (9): 2.

46. Mellano, M., Bounous, G. and Botta, G. 2005. Valutazione mediante analisi sensoriale di cultivar piemontesi di castagno. *Italus Hortus* 12(5): 64.

47. Zeppa, G., Rolle, L. and Gerbi, V. 2003. Studio per la caratterizzazione dei prodotti tradizionali regionali. *Il Marrone della Valle di Susa. Quaderni della Regione Piemonte* 38: 35–39

48. Graudal, N.A., Hubeck-Graudal, T. and Jürgens, G. 2012. Effects of low-sodium diet vs. high-sodium diet on blood pressure, renin, aldosterone, catecholamines, cholesterol, and triglyceride (Cochrane Review). *American Journal of Hypertension* 25(1): 1–15.

49. Chang, S.K., Alasalvar, C., Bolling, B.W. and Shahidi, F. 2016. Nuts and their co-products: The impact of processing (roasting) on phenolics, bioavailability, and health benefits – A comprehensive review. *Journal of Functional Foods* 26: 88–122.

50. Silva, A.P., Oliveira, I., Silva, M.E., Guedes, C.M., Borges, O., Magalhães, B. and Gonçalves, B. 2015. Starch characterization in seven raw, boiled and roasted chestnuts (*Castanea sativa* Mill.) cultivars from Portugal. *Journal of Food Science and Technology* 53(1): 348–358.

3 Distribution, Marketing, and Trade

Marta De Biaggi, Gabriele Beccaro, Jane Casey, Pedro Halçartegaray Riqué, Marco Conedera, José Gomes-Laranjo, Dennis W. Fulbright, Sogo Nishio, Ümit Serdar, Feng Zou, and Elvio Bellini

CONTENTS

3.1   CHESTNUT GLOBAL PRODUCTION

Chestnuts play an important role in diet, in the timber industry, and in land-scape design in many agri-economic systems worldwide. Over the past 30 years, the chestnut ecosystem has increased its ecological and landscape significance mainly through the planting of varieties resistant to fungal diseases and it has thus become a fundamental resource for the sustainable development of mountain areas.[1]

Indeed chestnut cultivation was affected by several problems that brought a dramatic collapse of the whole production system for years, in particular in western countries.

Most of the mountain communities in European countries were highly dependent on chestnut cultivation, both for timber and for fruits, harvested in great quantities (over 1 million t/year estimated at the beginning of the twentieth century).[2]

However, the appearance of new diseases (see Chapters 11 and 12) transported across chestnut commercial routes caused massive damages to *C. sativa* and *C. dentata* cultivations. Moreover, the advent of the industrial era and the substitution of more remunerative crops (potatoes, cereals, fodder, etc.) brought to a gradual abandonment of large chestnut growing areas.

Since 1990 the world chestnut production has been growing, reaching approximately 2.3 million tonnes of fresh fruits in 2017 over more than 600,000 ha.[3] East of Asia and Mediterranean Europe are the two main production areas, concentrating respectively almost 90% and over 7% of the chestnuts produced worldwide.

The Asian production is primarily focussed on *Castanea crenata* (Japanese chestnut), *C. mollissima* (Chinese chestnut), and their hybrids, while *C. sativa* traditionally dominates European chestnut cultivations, with growing interest for Eurojapanese hybrids. The main global producers are China (1.9 million tonnes in 2017, mainly *C. mollissima*), Turkey with more than 63,000 tonnes of *C. sativa*, Republic of Korea (53,000 t, mainly *C. crenata*), and Japan (18,700 t, mainly *C. crenata*). Italy is the second leading *C. sativa* producing country with about 50,000 tonnes, followed by Greece (36,000 t), Portugal (over 30,000 t), Spain (15,000 t), and France (8,000 t), as listed in Table 3.1.[3] Despite the lower productions compared to the above-mentioned countries, Australia, Chile, and the USA shall be considered expanding areas of production (Figure 3.1). FAO databases are updated quite regularly, but the chestnut market is difficult to quantify and data are often discordant: data from national experts and producers were included for more detailed information on trends at the national level.[4]

TABLE 3.1

Chestnut Production by Country

	Mean Value 1984/1988	% World	Mean Value 1994/1998	% World	Mean Value 2004/2008	% World	Mean Value 2011/2012	% World	2014	% World	2015	% World	2017	% World
China	96,018	21	345,000	47	1,162,243	77	1,706,244	83	1,689,735	82	1,668,913	82	1,939,716	83
Rep. Korea	66,222	15	108,359	15	76,677	5	63,466	3	59,465	3	55,593	3	52,764	2
Japan	49,420	11	31,300	4	23,260	2	20,000	1	21,400	1	16,300	1	18,700	1
Turkey	73,200	16	66,800	9	52,662	3	59,076	3	63,762	3	63,750	3	62,904	3
Italy	50,687	11	72,337	10	52,895	4	37,050	2	34,000	2	42,000	2	52,356	2
Greece	10,800	2	12,475	2	17,081	1	17,797	1	28,100	1	31,557	2	25,000	1
Portugal	17,476	4	25,124	3	27,501	2	18,701	1	18,464	1	27,628	1	29,875	1
Spain	30,349	7	13,274	2	9,856	1	16,100	1	16,136	1	16,413	1	15,623	1
France	15,300	3	11,000	1	8,632	1	8,600	<1	8,668	<1	7,700	<1	8,406	<1
World	452,178	100	734,018	100	1,506,361	100	2,049,584	100	2,057,019	100	2,044,428	100	2,327,495	100

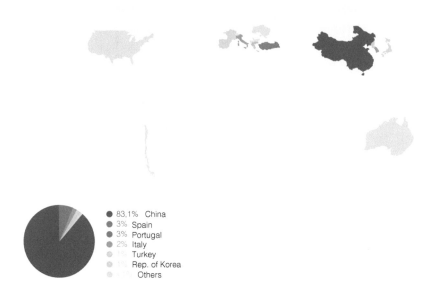

FIGURE 3.1   Map illustrating the main chestnut producing countries and relative market share.

3.2   GLOBAL TRADE SCENARIO

Even with lower volumes, in the last 40 years the chestnut in Europe has gradually began to regain ground, both for the cultivation of Eurojapanese hybrids and for the introduction of technological innovations in the timber and nut growing chain.[5]

Globally, over the past 15 years, the world chestnut production has been growing even with some noticeable fluctuations (Figure 3.2), mainly due to Chinese production increase. Indeed the intense scientific and agronomic activity promoted in the last decades of the twentieth century by the Chinese government provided a strong incentive for chestnut cultivation and cultivar improvement. Therefore, the impressive yield increase as a result of the national policies is responsible for the constant global production raise since 1990. Moreover, an increase in global chestnut production was driven by population growth and popularity of healthy eating. These key drivers are expected to continue promoting chestnut output in the immediate term.

Though the forest system area in Europe has been estimated at 2,000,000 ha, the area of agroforestry and crop systems is estimated at 300,000 ha. Chestnut production have been growing since 2010 in Turkey (+30%), Greece (+50%), and Portugal (+30%), while the production in France and Spain has been fluctuating.[3]

Italian production showed an increasing trend from 2007 to 2014 and a 5.6% reduction in 2014, followed by a recent slow production recovery.[3,6]

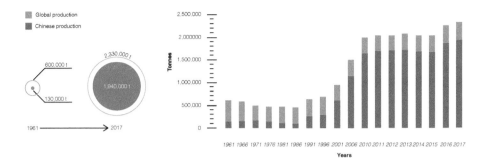

FIGURE 3.2 Global chestnut production trend (blue bars), and the contribution of the Chinese market (red bars) since 1961. (From FAOSTAT, *Food and Agriculture Organization of the United Nations Statistics Database*, Rome, Italy, 2018.)

3.3 PRODUCTION BY COUNTRY

3.3.1 TRADITIONAL CULTIVATION AREAS

3.3.1.1 China

China is the most important producer in Asia, with a long history of chestnut cultivation, being one of the oldest chestnut-growing countries. The Chinese chestnut is among the major cultivated plant species in China, and the chestnut industry is a crucial element in the country's economy (Figure 3.3). Being one of the most important edible crops in China, chestnuts are largely consumed across the country, both fresh and processed. Chestnut cultivations cover more than 300,000 ha[3] concentrated

FIGURE 3.3 Chestnut harvesting in China. (Courtesy of Zou, F.)

in the southeastern part of the country across 23 provinces and mainly along the Yangtze valley. Fruit production is based on *C. mollissima*, a species with high genetic variability, which easily cross-pollinates with other species such as *C. sativa* and *C. crenata* forming hybrids. More than 300 cultivars have been identified and divided in 6 regional groups.

3.3.1.2 Republic of Korea

While the Chinese chestnut is mainly present in North Korea, intensive farming of *C. crenata* and new hybrids have developed in the Republic of Korea. Following the destruction of the existing dense and extended forests caused by the gall wasp, in the 1970s a national genetic improvement program was implemented to select hybrids resistant to pests and to low temperatures. Nowadays intensive cultivations cover approximately 33,000 ha mainly concentrated in the Kwangyang, Kongju, Sanchong, Chungju, and Hadong provinces with large chestnut forest areas.[3,5,7] Approximately 30% of national production is consumed fresh.[8]

3.3.1.3 Turkey

C. sativa is one of the most common tree species in Turkey, with great ecological and economical importance for its timber, fruits, leaves, and flowers. There are very few studies on production and marketing characteristics of chestnut growers in Turkey. Only recently, Serdar et al. (2018) outlined production characteristics (germplasm, farms, pest and diseases problems and their management) and analyzed the Turkish market. The most important fruit production areas are concentrated in the Black Sea (30%), Marmara (9%), and Aegean regions (61%).[9] During the last decades of the twentieth century, fruit production slowly decreased due to *C. parasitica* attacks. While chestnut yield was 2.4 t/ha in the late 1980s, nowadays it is 1.5 t/ha.[10] However, Turkey still remains one of the top producers in the world, also as a result of a recent renewed interest in chestnut cultivation. Following the increase in the income of agricultural enterprises engaged in fruit farming, the rate of fruit cultivation area grew from 10.9% to 13.7% since 1990.[11]

Chestnut growers' income in the Aegean region comes more from chestnut wood and honey rather than from nuts. Growers in the East Black Sea region present a high total share of income from chestnut cultivation (18.3%), despite the fact that they only represent 3% of the total chestnut production in Turkey.[9]

3.3.1.4 Japan

Chestnuts are cultivated all over Japan, from Yakushima island to central Hokkaido except for the northern part of the island and Okinawa. Japan is one of the world's biggest chestnut consumers and mainly focuses on *C. crenata* with only a few hybrids *C. crenata × C. mollissima*, as the Chinese chestnut has low adaptation capacity to the country's pedoclimatic conditions. In the past, chestnut trees have been cultivated in hilly and mountain areas, but this cultivation area has steadily decreased due to the inconvenience of country life. Most of the new plantations are designed as fruit tree orchards on flat land. The major production areas are Ibaraki, Kumamoto, and Ehime Prefectures, which together account for 50% of the country's chestnut production.

FIGURE 3.4   Chestnut selection for commercialization in Japan. (Courtesy of Nishio, S.)

Despite the impressive recovery of chestnut cultivation after World War II (from 4,600 ha to 44,000 ha in 1975) as a consequence of government incentives and the selection of varieties resistant to the gall wasp, the Japanese chestnut market suffers from strong competition with China and Korea. Production volumes have indeed decreased in the last 20 years, and the cultivation surface has shrunk to about 20,000 ha (18,700 t in 2017)[3] (Figure 3.4).

Recently in Japan, more and more cooperatives and producers have been selling nuts after cold storage. It is also important for cooperatives and producers to control selling time, as the selling price temporarily declines in late September (the peak period of chestnut harvesting time), but chestnut demand continues into late December.

3.3.1.5   Italy

Italy is one of the largest *C. sativa* producing, consuming, and exporting countries. Chestnut-growing areas in Italy cover about 780,000 ha, 7.53% of the total forest area,[6] an important forest heritage, largely man-made, and mainly characterized by mixed forests and abandoned groves. Italy is also a leading country in producing processed chestnut products such as marron glacé. Despite the production decrease, a hard-core of growers remains managing approximately 50,000 ha. Chestnut production is concentrated in the central and southern regions. In particular, the prominent chestnut-producing regions are Campania (13,800 ha), Tuscany (10,400 ha), Calabria (8,600 ha), Piedmont (6,400 ha), Lazio (3,800 ha), and Emilia Romagna (2,800 ha). All other regions where chestnuts are grown contribute less than 10% to the national harvest.[4,12]

C. sativa and Eurojapanese hybrids cover almost 88% of total Italian production, whereas 12% is represented by marrone type varieties. Approximately 75% of the fresh chestnuts are consumed in local markets or exported, 20% is either dried or processed in the agro-food industry, and the remaining 5% is used as animal food.

3.3.1.6 Portugal

Thanks to European funding programs and the establishment of new plantations, Portuguese chestnut production has slowly been increasing after being strongly affected by ink disease, until a minimum area of 15,000 ha existed in 1970s. Actually, the orchards cover an area of about 35,000 ha,[3] producing 30,000 t in 2017 (Figure 3.5). It is noteworthy that according to RefCast, the Portuguese Association on Chestnut, the production is estimated around 45,000 tonnes. The main production areas are located between 400 and 1000 m a.s.l. in the interior part of the north and central regions, contributing Trás-os-Montes with 80% of the total production area, and Beira Interior with 10% of the total area. Small production areas can be found in the northeast part of Alentejo reaching the most western European regions, Minho (with significative increase in recent years), Madeira, and the Azores archipelago.[4]

3.3.1.7 Spain

The total area of chestnut in Spain is around 111,000 ha producing around 40,000 tonnes (Red Estatal del Castaño). The main chestnut producing areas in Spain are Galicia with 70,000 ha and Castilla y Leòn 29,000 ha, followed by Andalucía (9,000 ha), and Estremadura (3,000 t).[3] In Asturias, Cataluña, and Biscaye cultivations are present but are not significant. The renovation politics and the new plantations have proven a positive evolution of the yields in Galicia and Castilla y Leòn, from 1.5 t/ha to 3 t/ha;[4,13] however, future prospects still raise concerns.

FIGURE 3.5   Chestnut plantation in Portugal. (Courtesy of Beccaro, G.)

3.3.1.8   Greece

C. sativa occurs from the Greek Northern borders down to Crete Island covering approximately 10,000 ha, with higher concentrations in Thessalia (15.3% on Mount Pilion), Mount Athos (23.5%), and Thrace, at altitudes between 300–1100 m a.s.l.[3,5] Greek production has been growing in the last ten years, reaching 25,000 tonnes in 2016 (S. Diamandis 2018, pers. comm.). However, the presence of chestnut rot caused by *Gnomoniopsis castanea*, probably as a consequence of climate anomalies during the spring season (unusually heavy rains), is raising serious concerns regarding future yields (S. Diamandis 2018, pers. comm.).

3.3.1.9   France

Currently, the main chestnut-growing areas in France cover less than 9,000 ha and are located in southeastern regions (Ardèche accounting for 46%, Cévenne and Languedoc-Roussillon-Tarn), southwestern regions (Dordogne-Bouriane accounting for 10%, Limousin and South Massif Central), and Corsica (5%) with several secondary producing areas such as the Pyrenees region, Maritime Alps, and Bretagne.[3,5] Production levels have been constant since the beginning of the twenty-first century. Several initiatives across the main production areas such as Var, Haute Provence, and Pyrenees are focusing on the recovery of abandoned groves and the valorization of the existing production chain.[5]

3.3.1.10   Switzerland

In Switzerland *C. sativa* is the most abundant species in terms of tree number, covering 15% of the total forest area (27,000 ha)[5] (Figure 3.6). In the last century, chestnut cultivation has been highly affected by socio-economic changes and the introduction

FIGURE 3.6   Agroforestry in Swiss chestnut grove. (Courtesy of Beccaro, G.)

of chestnut blight. Only recently, a renewed interest in the cultivation of sweet chestnut is helping to revive the species cultivation for timber production and multipurpose management, improving chestnut production (150 t in 2017).[3,14]

3.3.1.11   Continental Europe

In most of the central European countries, accounting for less than 11% of *C. sativa* area, there has only been a partial development of a chestnut tradition, due to changes in climate conditions in certain areas (Germany, Austrian inner-Alpine regions, Romania, Czech Republic, Poland, and Ukraine), their particular geography (e.g., England), their history (e.g., Slovenia, Croatia, Georgia), the sporadic occurrence of the species (e.g., Hungary, Belgium, Bulgaria), or for its recent introduction (e.g., Slovakia and the Netherlands).[15]

However, chestnut coppices and orchards (not always in good conditions) are grown across the Caucasus, Carpathians, and the Balkans, generally with very low fruit production: approximately 70,000 ha in Georgia, 3,500 ha in Kosovo, less than 3,000 ha across Romania; 15,000 ha of *C. sativa* are spread over hilly and mountainous areas in Albania, 135,000 ha in Croatia, extremely limited areas in Southern Slovakia (130 ha), and Slovenia (32 ha).[16,17] Bulgaria, Hungary, and Poland showed similar productions in 2017 with approximately 500 tonnes. In Germany sweet chestnut is spread in the Southern part of the country, covering approximately 7,500 ha.[18]

3.3.2   EMERGING COUNTRIES

3.3.2.1   United States

Chestnut production in the United States is still not a large business but is expected to continue to increase. Within the first half of the twentieth century, after the chestnut blight began devastating the American chestnut (*C. dentata*) tree population in its natural range, researchers discovered that *C. mollissima* was resistant to chestnut blight. The Chinese chestnut rapidly gained popularity in the Midwest[19] and the Appalachian Mountain states to replace nut production lost with the loss of the American chestnut trees.

Today, the largest and most successful Chinese chestnut farms in the Appalachian Mountain territory are found in Pennsylvania, Ohio, North Carolina, and Florida. In the Midwest, growers in states like Illinois, Iowa, Missouri, Kansas, and Oklahoma established Chinese chestnut seedling and grafted tree orchards.

In Western states such as California, Oregon, and Washington state where chestnut blight had a milder impact, when it came to farm establishment, the growers generally started by planting grafted cultivars of *C. sativa* × *C. crenata*, or by planting rootstock and grafting *C. sativa* × *C. crenata* hybrid cultivars.

The top five states with the most chestnut acreage are Michigan, Florida, California, Oregon, and Virginia together with Pennsylvania and Ohio.[5]

Michigan plants more chestnut cultivars in orchards than any other state, and chestnut production is estimated around 110,000 kg chestnuts in most years. However, as not all growers belong to cooperatives and many producers offer their harvest for sale directly to consumers or farm by farm, it is problematic to assess the actual production and sales in the United States chestnut market.[8] U.S. chestnut production is less than 1% of total world production, covering over 3,700 ha, according to the Agricultural Marketing Resource Center. However, recent genetic research is concentrated on the improvement of cultivars resistant to pests and pathogens and the restoration of the *C. dentata* on the Appalachian Mountains, its original growing territory, proving the growing interest in chestnut products.

3.3.2.2 Chile

Similarly to the United States, in Chile, despite being a minor crop, chestnut cultivation is recently receiving more attention (Figure 3.7). Chile has the best competitive access to the off-season chestnut market in the northern hemisphere. Chile is the biggest nut exporter in the southern hemisphere and possesses various ideal conditions for growing chestnut trees: fertile soils, low risk of spring frosts, high humidity, and no reports of *C. parasitica*. The average yield of chestnut orchards in Chile is about 5,000 kg/ha.[20] Indeed, a few years ago there were almost no commercial orchards, simply some groups of seedling trees planted near the houses at the growers' properties. For this reason there is still little information available on the crop.

3.3.2.3 Australia and New Zealand

The chestnut industry in Australia is cantered in North East Victoria, accounting for some 75% of the total Australian production (Figure 3.8). Other growing regions include Batlow/Tumbarumba and the Blue Mountains in New South Wales, the Adelaide Hills in South Australia, and around Manjimup in Western Australia. The total crop is estimated to be around 1,500 tonnes, sold domestically as fresh chestnuts.

In New Zealand the chestnut industry is less than 20 years old, coordinated by the NZ Chestnut Council since 2000. Chestnuts grow successfully across most of New Zealand, with higher orchard concentrations in the Waikato, Bay of Plenty, and Auckland areas. The most commonly planted varieties are the Eurojapanese hybrids and Japanese chestnut varieties. As of 2016, fresh nut production was around 300–400 tonnes (NZ Horticulture Export Authority).

FIGURE 3.7   Chestnut plantation in Chile. (Courtesy of Halçartegaray, P.)

FIGURE 3.8   Chestnut plantation in Australia. (Courtesy of Griffiths, S.)

3.4   CHESTNUT EXPORT AND IMPORT TRENDS

3.4.1   GLOBAL EXPORT TREND

Overall, global chestnut export indicated a pronounced growth from 2007 to 2016 (+2.6% per year), amounting to $349 million in 2016, with noticeable fluctuations.[3] China is the world's top chestnut exporter, with over 27% of the total exports, followed by Spain and Portugal both 14%, and Italy (11%), together accounting for 52% of total exports in 2016 as shown in Figure 3.9.[3] Spain showed notable growth rates of chestnut exports compared to their global leaders, moving from 5,600 tonnes in 2008 to over 20,000 tonnes in 2016.[3] Portugal as well shows a growing trend since 2008, from 7,800 tonnes to 19,800, traditionally with large size nut mainly exported to Brazil (around 1,800 t). Recently, exportations to Europe significantly increased; exports in 2012 included Italy (4,400 t), France (4,400 t), and Spain (2,200 t).

The data on Italian chestnut trading activities demonstrate the national production difficulties. Italy has always been an exporting country, and its fresh and processed products are highly appreciated and imported by Germany (22%), Switzerland (18%), France (17%), Austria (13%), and the United States (8.5%). Since 2006, however, a drastic reduction in export volumes brought an increase in imported nuts, and from 2010 to 2016 the export situation worsened.[3] Despite the market difficulties due to the severe phytosanitary emergencies, Italy remains, together with China, one of the main actors on international markets for the chestnut export value thanks to the remarkable organoleptic and aesthetic properties of its cultivars. Italian exports represent 11% of the volume and 21% of the value of exported chestnuts in the world, whilst Chinese exports represent 31% of the volume and 22% of the global value of exported chestnuts.[6] In particular, in Italy,

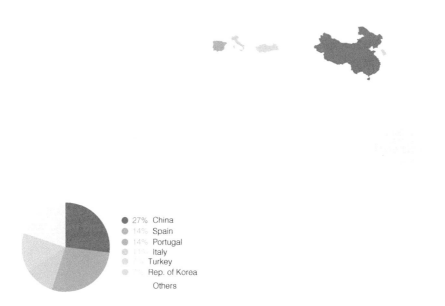

FIGURE 3.9 Chestnut top exporters. (From FAOSTAT, Food and Agriculture Organization of the United Nations statistics database, Rome, Italy, 2018.)

Campania is the main exporting region (40% of total exported volumes), followed by Piedmont (16%) (CREA, Banca dati Commercio Estero). However, a decrease in exported volumes from Campania has occurred in the last 5 years, which on the contrary has increased in Piedmont.

In the last ten years, Turkey presented an overall increase in chestnut exports (from less than 1,000 t in 2007 to 8,300 t in 2016), with fluctuations throughout the analyzed period.

Despite the efforts and the great improvement in quality and production levels, the Chinese market is strongly influencing Korean exports, especially towards Japan. For this reason, since 2008, the Republic of Korea started an exportation program towards European countries.[4]

Only a small amount of fresh Australian chestnuts is exported, mainly to Southeast Asia, and some small scale processing is undertaken. However, this accounts for such small volumes to be negligible.

Exports from Chile occur from May to September because harvest begins in April. Volumes and prices are clearly defined by the availability of European chestnuts. The interest to buy Chilean chestnuts is increased on low European crop years. The main destinations for exported chestnuts are France, Italy, Spain, and Portugal and almost exclusively to the industries. In Chile, intensive *C. sativa* new plantations are made by agricultural companies that also directly manage marketing through processing and export companies. The volume is expected to rise to 5,000 tonnes of marrone type varieties in the next four years.

3.4.2 GLOBAL IMPORT TREND

Globally, chestnut import volumes show a relatively flat trend, reaching a maximum in 2014 with 125,000 tonnes and slightly decreasing in the following years. In value terms, chestnut imports amounted to $343 million in 2016.[3] Italy is the top importing country (37,000 t), followed by China (13,500 t), France (8,700 t), Japan (8,000 t), Thailand (5,200 t) and Taiwan (4,700 t), Germany (3,400 t), the United States (3,200 t), Lebanon (3,100 t), Switzerland (2,500 t), Republic of Korea (2,300 t), and Israel (1,800 t) according to 2016 Faostat data[21] (Figure 3.10).

As previously discussed, while Italy is suffering a decrease in chestnut exports, imports increased by 74% reaching 37,000 tonnes in 2016. Italy alone accounts for 25% of the global import value ($64 million), followed by China with a 9% share of global imports, and Germany (7%).[22] Italy's main suppliers are Spain (33%), Portugal (23%), and Turkey (14%), followed by Greece. Together with the Italian market, France, Taiwan, Germany, Israel, Thailand, and Republic of Korea displayed positive import trends. In particular France showed the fastest growth trend in the world (+13.54% from 2007 to 2016). Spain, Portugal, and Italy are France's main suppliers. Turkey has also increased chestnut import quantity from 78 to 206 tonnes during the last two decades,[10] with a considerable decrease in 2016.[3] In Figure 3.11 are represented the main chestnut producers and relative import and export flows discussed in this chapter.

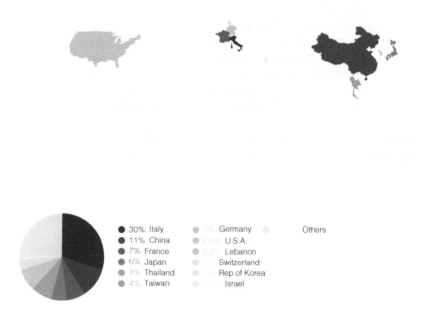

FIGURE 3.10   Chestnut top importers. (From FAOSTAT, Food and Agriculture Organization of the United Nations statistics database, Rome, Italy, 2018.)

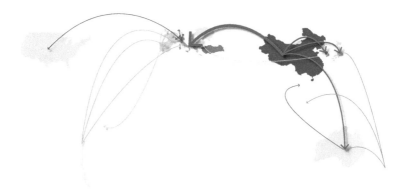

FIGURE 3.11 Chestnut top producers and relative import and export flows. (From FAOSTAT, Food and Agriculture Organization of the United Nations statistics database, Rome, Italy, 2018.)

The impressive amounts of chestnuts imported and exported described above demonstrate the importance of this fruit as a food, also given by its enormous versatility. Indeed a wide range of processed items can be obtained from chestnuts and have been appreciated for centuries all over the world. Table 3.2 shows some of the main dishes prepared with chestnuts (Figure 3.12).

TABLE 3.2

Recipes and Specialities Cooked with Chestnuts

Appetizer/Entree	Roasted chestnuts; chestnut ricotta and honey cream; chestnut hummus dip; chestnut soufflé; bread croutons with marroni, hazelnuts, and truffle; boiled chestnuts with milk
Main course	Chestnut soups with pumpkin and pancetta or with other vegetables; marroni cream; chestnut gnocchi, ravioli, or tortelli; chestnut and mushroom risotto
Second course	Roast or braised veal and chestnuts; roast pork with apple and chestnuts; game with chestnut sauce; chestnut polenta; roasted Asian-style chicken with chestnuts; salads with chestnuts; chestnut bread
Dessert	Marrons glacés; marrons glacés cake; chestnut Mont Blanc; creamy chestnut tiramisu; chestnut jam, honey, pudding, cookies, fritter, ice-cream, pastries, and buns
Drinks and liquors	Chestnut beer, liquor, and Amarone

FIGURE 3.12   Chestnut consumption presents a growing trend in Thailand. (Courtesy of Beccaro, G.)

3.5   CHESTNUT STAKEHOLDERS

In this section are listed some national stakeholders (enterprises, associations, and cooperatives) that promote chestnut production and trade (the list is indicative, not exhaustive).

Australia
   Chestnuts Australia Inc.
Chile
   Vivero Austral
   Chilean Marroni Farms (CMFexport)
   SubSole Nuts
   Agroindustrias San Francisco
China
   National Innovation Alliance of Chestnut Industry
   Non-wood Forest Branch of Chinese Society of Forestry
   Dried Fruit Branch of Chinese Society for Horticultural Science
   Chestnut Branch of China Cash Forest Association
Europe
   Eurocastanea—European Chestnut Network
France
   Chambre d'agriculture de l'Ardèche
   Comité Interprofessionnel de la Chataîgne d'Ardèche
   Union Interprofessionnelle de la Châtaigne Périgord-Limousin-Midi
      Pyrénées

Greece

Cooperation of Fruit Production & Trade Ormas Almopias

Agricultural Cooperative Paikou

Agricultural Cooperative of Olive & Chestnut Producers' Paleon Roumaton

Agricultural Cooperative of Producers of Melivoia

Agricultural Cooperative Potamias & Skitis

Agricultural Cooperative of Karitsa & Stomio

Italy

Associazione Nazionale Città del Castagno

Centro Studio e Documentazione sul Castagno, Marradi

Chestnut R&D Centre Piemonte

Japan

Ja-Zenchu (Central Union of Agricultural Co-operatives), includes several
small cooperatives

New Zealand

New Zealand Chestnut Council

Portugal

RefCast: Portuguese Association on Chestnut

Spain

Centro de Servicios y Promoción Forestal y de su Industria de Castilla y Leon

Andalousie: Junta de Andalucia, Séville, www.juntadeandalucia.es

Catalogne: Generalitat de Catalunya, Barcelona, www.gencat.cat

Valencia: Generalitat Valenciana, www.gva.es

IGP Galicia

Red Estatal del Castaño

Switzerland

Associazione dei Castanicoltori della Svizzera italiana

IG Pro Kastanie Zentralschweiz

Confrérie des amis de la châtaigne

Turkey

Chestnut Research Group, Samsun

United States

Chestnut Growers Inc.

Agricultural Marketing Resource Center

Michigan Nut Growers Association

American Chestnut Foundation

REFERENCES

1. Vollmeier, R., Osterc, G. and Luthar, Z. 2018. Preservation of sweet chestnut genetic resources (*Castanea sativa* Mill.) against attack by chestnut gall wasp (*Dryocosmus kuriphilus* Yasumatsu, 1951). *Acta agriculturae Slovenica* 111(1): 209–217.
2. Avanzato, D. and Bounous, G. 2009. Following chestnut footprints *Castanea* spp. *Scripta Horticulturae* 9: 1–175.
3. FAOSTAT. 2018. *Food and Agriculture Organization of the United Nations Statistics Database*, Rome, Italy.

4. A.R.E.F.L.H. 2017. *Livre Blanc de la Chataigne Europeenne.* AREFLH, Bordeaux, p. 39.
5. Bounous, G. 2014. *Il castagno: Risorsa Multifunzionale in Italia e nel Mondo.* Edagricole, Bologna, p. 420.
6. MIPAAFT, 2018. Piano di settore castanicolo. https://www.politicheagricole.it/flex/cm/pages/ServeBLOB.php/L/IT/IDPagina/3277.
7. Kim, M., Lee, U., Kim, S., Hwang, M. and Lee, M. 2004. Comparison of nut characteristics between Korean native chestnut accessions and prevailing cultivars cultivated in Korea. *Acta Hortic.* 693: 299–304.
8. Metaxas, A.M. 2013. Chestnut (*Castanea* spp.) cultivar evaluation for commercial chestnut production in Hamilton County, Chattanooga, TN.
9. Serdar, Ü., Akyüz, B., Ceyhan, V., Hazneci, K., Mert, C., Er, E., Ertan, E., Savaş, K.S.Ç. and Uylaşer, V. 2018. Horticultural characteristics of chestnut growing in Turkey. *Erwerbs-Obstbau* 60(3): 239–245.
10. Bozoglu, M., Baser, U., Eroglu, N.A. and Topuz, B.K. 2017. Developments in the chestnut market of Turkey. *Proceedings of the 6th International Chestnut Symposium, Samsun (Turkey), Acta Horticulturae* 1220, Leuven .
11. Güneş, N.T., Horzum, Ö. and Güneş, E. 2017. Economic and technical evaluation of fruit sector in Turkey. *Balkan and Near Eastern Journal of Social Sciences* 3(2): 37–49.
12. Pierrettori, S. and Venzi, L. 2009. The chestnuts Filiere in Italy: Values and developments. *EFI Proceedings* 57: 85–96.
13. Pereira-Lorenzo, S., Díaz-Hernández, M.B. and Ramos-Cabrer, A.M. 2009. Spain: In following chestnut footprints (*Castanea* spp.): Cultivation and culture, folklore and history, traditions and uses. *International Society of Horticultural Science, Scripta Horticulturae* 9: 134–142.
14. Conedera, M. and Krebs, P. 2009. Switzerland: In following chestnut footprints (*Castanea* spp.): Cultivation and culture, folklore and history, traditions and uses. *International Society of Horticultural Science, Scripta Horticulturae* 9: 149–154.
15. Conedera, M., Manetti, M., Giudici, F. and Amorini, E. 2004b. Distribution and economic potential of the Sweet chestnut (*Castanea sativa* Mill.) in Europe. *Ecologia Mediterranea* 30(2): 179–193.
16. Tahiri, V. 2018. *The Chestnut Market and Consumption in Kosovo.* Academic Journal of Business, Administration, Law and Social Sciences IIPCCL Publishing, Graz-Austria Vol. 4, No. 3
17. Kos, K., Kriston, E. and Melika, G. 2015. Invasive chestnut gall wasp Dryocosmus kuriphilus (Hymenoptera: Cynipidae), its native parasitoid community and association with oak gall wasps in Slovenia. *European Journal of Entomology* 112(4): 698.
18. Bouffier, V.A. and Maurer, W.D. 2009. Germany: In following chestnut footprints (*Castanea* spp.): Cultivation and culture, folklore and history, traditions and uses. *International Society of Horticultural Science, Scripta Horticulturae* 9: 53–62.
19. Warmund, M.R. 2011. Chinese chestnut (*Castanea mollissima*) as a niche crop in the central region of the United States. *HortScience* 46(3): 345–347.
20. Joublan, J.P., Ríos, D. and Montigaud, J.C. 2004. The competitive advantage of Chilean national fresh chestnut industry. *III International Chestnut Congress* 693: 55–62.
21. FAOSTAT. 2016. *Food and Agriculture Organization of the United Nations Statistics Database.* Rome, Italy.
22. IndexBox. 2018. World—Chestnut—Market analysis, forecast, size, trends and insights. www.indexbox.io.

4 Cultivars List and Breeding

CONTENTS

4.1 CULTIVARS LIST (FIGURES 4.1 THROUGH 4.6)

Isidoro Riondato, Burak Akyüz, Gabriele Beccaro, Jane Casey,
Marco Conedera, Jean Coulié, Stephanos Diamandis,
José Gomes-Laranjo, Sogo Nishio, Santiago Pereira-Lorenzo,
Ana Ramos-Cabrer, Ümit Serdar, Feng Zou, and Michele Warmund

Cultivar	Species	Synonyms	Origin	Nut Size	Ripening Time	Pellicle Adhesion	% of Doubles	Nut Stipes	Catkins	Remarks/Main Utilizations
Abadá	S	–	Spain	L	Late–very late	Complete	Low	Thin	M	Flour, marmalade, purée
Abarcá	S	–	Spain	S	Late–very late	Variable	Nul	Thin	M	Fresh, flour, marmalade, purée
Aguyane	S	–	France	M–L	–	–	–	–	–	Fresh
Akachiu	C	Akatyu	Japan	L	Medium	Complete	Low	–	M	–
Akyüz	Y	–	Turkey	L	Very early	Free	Very high	Very thin	L	On registration process/I(–)
Alachua	M × D	–	USA	–	–	–	–	–	–	–
Ali Nihat	Y	–	Turkey	L	Very early	Free	Very high	Very thin	L	On registration process/I(–)
Allegheny chinkapin	P	–	USA	XS	Very early	–	–	–	–	C. pumila var. pumila/ Ornamental, nut
Amadengue	S	–	Spain	XS	Late–very late	Free	High	Very thin	–	Flour, marmalade, purée
Amarelal	S	–	Portugal	L	Early	Free	Low	–	–	–
Amarelante	S	Marela, Amarela	Spain	M	Late–very late	Partial	Low	Thin	M	Different genotypes/Fresh, flour, marmalade, purée
American chestnut	D	–	North America	XS	–	Free	–	–	–	Timber
Amy	M	–	USA	M	Early	–	–	–	–	–
Anaxa	S	–	Spain	M	Late	Partial	Nul	Thick	M	Fresh, flour, marmalade, purée

(Continued)

Cultivar	Species	Synonyms	Origin	Nut Size	Ripening Time	Pellicle Adhesion	% of Doubles	Nut Stipes	Catkins	Remarks/Main Utilizations
Arafero	S	–	Spain	L	Late–very late	Partial	High	Very thin	M	Marron glacé
Arcadia	S	Pamona, Tripoli	Greece	L–XL	Early–medium	Free	Low	–	–	Fresh, candied, confectionary
Argua	S	Argua Roxa	Spain	M	Late–very late	Free	Moderate	Very thin	L	Fresh
Arial	S	–	Spain	M	Late–very late	Free	Low	Thick	L	Fresh, flour, marmalade, purée
Arima	C	–	Japan	L	Early–medium	Complete	–	–	–	–
Armstrong	M × D	–	USA	L	–	–	–	–	–	–
AU-Cropper	M	–	USA	S	–	–	–	–	–	–
AU-Homestead	M	–	USA	M	Late	–	–	–	–	Fresh, roasted
AU-Leader	M	–	USA	L	–	–	–	–	–	–
AU-Super	M	–	USA	XL	–	–	–	–	–	–
Aveleira	S	–	Portugal	M	Early	Free	Nul	Thin	–	–
Azirinca	S	–	France	M	Late	–	Nul–low	–	–	Fresh, processed
Bacoa	S	Bacón	Spain	L	Medium	Free–Partial	High	Thin–very thin	L	Fresh, marron glacé
Banseki	C	–	Japan	XL	Medium–late	Complete	Moderate	–	L	–
Baopidayouli	M	–	China	L	Early	–	–	–	–	Cooked, boiled
Baria	S	–	Portugal	L	Medium	Free	Low	Thin	–	–
Bebim	S	–	Portugal	L	Medium	Free	Low	Thick	–	–
Belle Epine	S	–	France	L	Medium–late	Partial	Low–moderate	Thin	L	Pollinizer, I(–)/ Fresh, processed

(Continued)

Cultivar	Species	Synonyms	Origin	Nut Size	Ripening Time	Pellicle Adhesion	% of Doubles	Nut Stipes	Catkins	Remarks/Main Utilizations
Bellefer®	S × C	–	France	–	–	Free	–	–	–	Fruit
Benfeita	S	–	Portugal	M	Medium	Free	Nul	Thick	–	–
Benton Harbor	M	–	USA	–	–	–	–	–	–	–
Berciana	S	–	Spain	S	–	Free	Low	–	–	Flour, marmalade, purée
Bermella	S	Bermellas, Bermellá, Bermellal	Spain	M	Late–very late	Free	Nul	Thin	M	Different genotypes/Fresh, marmalade, purée
Beth	M	–	USA	–	–	–	–	–	–	–
Bicuda (Azores)	S	–	Portugal	S	Late	–	–	–	–	–
Bionda di Mercogliano	S	Ionnola	Italy	M	Medium	–	–	–	–	Fresh
Bisalta#3	S × C	–	USA	L	–	–	–	–	–	Pollinizer/Fresh, processed
Blanca	S	Pallarega	Spain	S	Late–very late	Partial	Nul	Thin	A	Different genotypes/Fresh, flour, marmalade, purée
Boaventura	S	–	Portugal	M	Late	Free	Nul	Thick	–	–
Bolesas	S	–	Spain	S	Late–very late	Partial	Nul	Thick	M	Pollinizer/Marmalade, purée
Bonguri	C	–	Japan	–	–	Complete	–	–	–	–
Bonosora	S	Bonosole, Bonosola	Italy	M	Medium	Partial	–	–	–	I(+)/Fresh

(Continued)

Cultivar	Species	Synonyms	Origin	Nut Size	Ripening Time	Pellicle Adhesion	% of Doubles	Nut Stipes	Catkins	Remarks/Main Utilizations
Boroñona	S	Baragaña	Spain	S	Medium–late	Partial	Low	Very thin	L	Flour, marmalade, purée
Bouche de Bétizac	S × C	–	France	L–XL	Very early–early	Free	High	Thick	A	Worldwide distributed/I(–)/Fresh, processed
Bouche Rouge	S	Marron de Vesseaux, Grosse Bouche, Bouche	France	M–L	Very late	Partial	Nul-low	Thick	A	I(–)/Fresh, processed
Bournette	C × S	–	France	L–XL	Early–medium	Free–Partial	Low	Thick	L	Worldwide distributed/Fresh
Bracalla	S	–	Italy	M	Medium	Partial	Very high	Thin	A	I(–)/Fresh
Bravo de Leirado	S	–	Spain	S	Late–very late	Free	Low	Very thin	L	Pollinizer/Fresh, candied, flour, marmalade, purée
Brulina	S	–	Spain	S	Late	Free	Nul	Very thin	L	Flour, marmalade, purée
Brunette	S	–	Italy		Medium	–	–	Thin		Fresh, flour, marmalade, purée
Buffalo Queen	–	–	Australia	–	–	–	–	–	–	–
Burgaceira	S	Burgueceira	Spain	M	Late–very late	Free	Low	Thick	A	Pollinizer/Fresh, candied, flour, marmalade, purée
Buzen	C	–	Japan	XL	Medium	Complete	Low	–	L	Low
Byron	M	Lindstrom # 67	USA	L	–	–	–	–	–	–
Cabezuda	S	Temporá, Cedo	Spain	L	Late	Partial	Low	Thin	M	Fresh, flour, marmalade, purée
										(Continued)

Cultivar	Species	Synonyms	Origin	Nut Size	Ripening Time	Pellicle Adhesion	% of Doubles	Nut Stipes	Catkins	Remarks/Main Utilizations
Calva Asturias	S	–	Spain	S	Medium–late	Free	Very high	Very thin	L	Flour, marmalade, purée
Calva Galicia	S	Calvas	Spain	M	Late–very late	Partial	Low	Thin	M	Flour, marmalade, purée
Camberoune	S	–	France	M-L	Medium–late	Free	Nul–low	Thin	–	Fresh
Campbell NC-8	M	–	USA	M	–	–	–	–	–	–
Campilla	S	Campillo	Spain	XS	Medium–late	Free	Nul	Thin	M	Flour, marmalade, purée
Canby Black	S	–	USA	–	–	–	–	–	–	–
Canepina	S	–	Italy	M	Medium–late	Partial	Low	Thin	M	Fresh, marmalade, purée
Capannaccia	S	Capannacce, Crepola, Capannaccina	Italy	M	Medium	Partial	–	–	–	Fresh
Capilla	S	-	Spain	L	Medium–late	Partial	High	Thin	L	Fresh
Capranica	S	Prenestina	Italy	M	Medium	Partial	–	–	–	Fresh
Caranquexa	S	–	Spain	M	Medium–late	Complete	Low	Very thin	L	Fresh
Carolina	M × D	–	USA	–	–	–	–	–	–	–
Carpinese	S	Carpanese, Carrarese	Italy	S	Medium	Partial	–	–	–	Fresh, flour
Carr	M	–	USA	M-L	–	–	–	–	–	–
Carreiró	S	–	Portugal	S	Medium	Free	–	–	–	–

(Continued)

Cultivar	Species	Synonyms	Origin	Nut Size	Ripening Time	Pellicle Adhesion	% of Doubles	Nut Stipes	Catkins	Remarks/Main Utilizations
Carrelaos	S	Carrelau	Spain	M	Late	Free	Low	–	M	Pollinizer/Fresh, candied, flour, marmalade, purée
Castagna di Vallerano	S	–	Italy	L	Medium	Partial	–	–	–	I(−)/Fresh, processed
Castagrande	S	Grande	Spain	S	Late	Partial	Moderate	Thin	L	Flour, marmalade, purée
Castelás	S	Castellanas	Spain	–	–	–	–	–	M	
Caurelal	S	–	Spain	XS	Late–very late	Free	High	Very thin	B	Flour, marmalade, purée
Cece dell'Amiata	S	–	Italy	M	Medium	Partial	–	–	–	I(−)/Fresh, processed
Cecio	S	–	Italy	M	Early	Partial	–	–	–	I(+)/Fresh
Cerreda	S	Cerredo	Spain	M	–	Complete	Low	Thin–very thin	M	Fresh, flour, marmalade, purée
Cesarucca	S	Cesarucche, Cesarucco	Italy	S	Medium–late	Partial	–	–	–	Fresh
Chamberga or Valduna	S	Valcuía, Valdín, Valduno, Marimoeches	Spain	M	Medium–late	Partial	Low	Thin	L	Fresh, flour, marmalade, purée
Chancloya	S	Chancloia, Forniega, Fornera, Marniega, Doriga	Spain	S	Medium–late	Free	Low	Thin	A	Flour, marmalade, purée
Changantiedanli	M	–	China	S	late	Free	–	–	–	Cooked, roasted *(Continued)*

Cultivar	Species	Synonyms	Origin	Nut Size	Ripening Time	Pellicle Adhesion	% of Doubles	Nut Stipes	Catkins	Remarks/Main Utilizations
Chenguoyouli	M	–	China	M–L	Early	Free	Low	–	–	Cooked
Chili	M	–	China	L	Very early–early	–	–	–	–	Cooked
Chocho	S	Chocho negro	Spain	M	Late	Free	Moderate	Thin–very thin	L	Fresh
Choubei	C	–	Japan	XL	Medium–late	Complete	Moderate	–	L	–
Choukouji	C	–	Japan	XL	Medium–late	Complete	Moderate	–	M	–
Chushuhong	M	–	China	L	Very early–early	Free	Low	–	–	Cooked
Ciapastra	S	–	Italy	M	Medium–late	Partial	–	–	–	I(+)/Fresh, flour
Ciuffa	S	–	Italy	S	Medium	–	–	–	–	Fresh
Colarinha	S	–	Portugal	M	Late	Free	Nul	Thin	–	–
Colossal	S×C	–	USA	XL	Early	Free–Partial	Moderate	–	A	–
Colunga	S	Galliciana, Zapatona	Spain	M	Late	–	Moderate	Thin–very thin	L	Fresh, candied
Colutad®	S×C	–	Portugal	M	Early	Free	–	Thin–thick	–	Rootstock
Comballe	S	Combasle	France	M–L	Medium–late	Free	High	Thick	–	Fresh, processed
Comisaria	S	–	Spain	L	Late	Partial	Low	Thin–very thin	L	Fresh, candied
Contessa	S	–	Italy	M	Early–medium	Partial	–	–	L	I(+)/Fresh

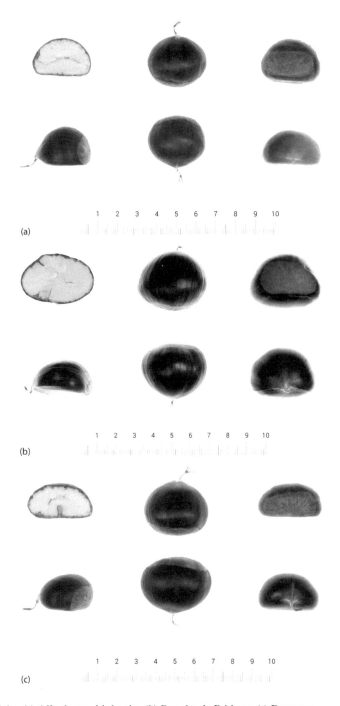

FIGURE 4.1 (a) Allegheny chinkapin; (b) Bouche de Bétizac; (c) Brunette.

(*Continued*)

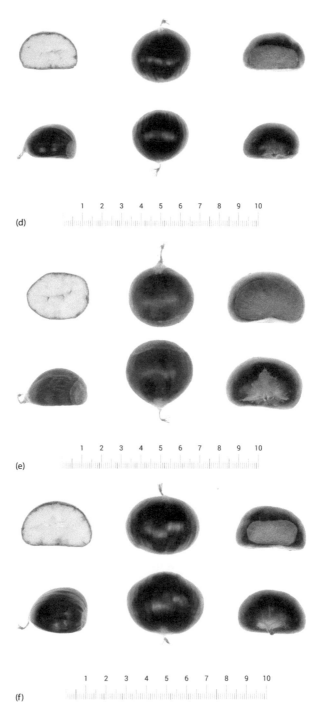

FIGURE 4.1 (Continued)   (d) Canepina; (e) Colossal; and (f) Contessa. (a, c, d, e, f: Courtesy of Betta, M.; b: Courtesy of Gamba, G.)

Cultivar	Species	Synonyms	Origin	Nut Size	Ripening Time	Pellicle Adhesion	% of Doubles	Nut Stipes	Catkins	Remarks/ Main Utilizations
Coração de boi (Madeira)	S	–	Portugal	M	Late	–	–	–	–	–
Corriente	S	–	Spain	L–XL	Very early–early	Partial	Very high	Variable	L	Different genotypes/ Fresh
Corujero	S	–	Spain	XS	–	–	Nul	Thin	L	Flour, marmalade, purée
Côta	S	–	Portugal	M	Late	Free	Nul	Thin–thick	–	–
Crane	M	–	USA	L	–	–	–	–	–	–
Crespa	S	–	Spain	XS	Medium–late	Free	Low	Very thin	L	Flour, marmalade, purée
Cruz	S	–	Spain	S	Late	Free	Very high	Thin–very thin	B	Flour, marmalade, purée
Cuijabaozi	M	–	China	S	Early–medium	Free	–	–	–	Roasted
Culochico	S	–	Spain	S	Very late	Free	High	Thick	L	Flour, marmalade, purée
Culona	S	Desgrañadiza	Spain	M	Late–very late	Partial	Low	Thick	M	Fresh, flour, marmalade, purée

(Continued)

Cultivar	Species	Synonyms	Origin	Nut Size	Ripening Time	Pellicle Adhesion	% of Doubles	Nut Stipes	Catkins	Remarks/ Main Utilizations
Curcia	S	–	Italy	S	Early–medium	Partial	–	–	–	Fresh
Curral (Madeira)	S	–	Portugal	M	Medium	–	–	–	–	–
Curuxa	S	Parede	Spain	XS	Late–very late	Free	Low	Very thin	M	Flour, marmalade, purée
Da Lebre	S	–	Spain	M	Late–very late	Free	Nul	Thin–very thin	B	Fresh, candied, flour, marmalade, purée
Dabanhong	M	-	China	S	Very early–early	–	–	–	–	Cooked, roasted
Dadiqing	M	–	China	XL	Early	Free	Low	–	–	Cooked
Daebo	C	–	South Korea	XL	Early	Complete	–	–	–	–
Dahongpao	M	–	China	L	Early	Free	Low	–	–	Cooked
Daihachi	C	–	Japan	XL	Medium–late	Complete	Low	–	M	–
Daiyuezaofeng	M	–	China	M–L	Very early	–	–	–	–	Cooked
Das Anchas	S	–	Spain	S	Late	Free	Nul	Thin	M	Fresh, flour, marmalade, purée

(Continued)

Cultivar	Species	Synonyms	Origin	Nut Size	Ripening Time	Pellicle Adhesion	% of Doubles	Nut Stipes	Catkins	Remarks/ Main Utilizations
Dauphine	S	Marron Dauphine, Dauphinenque	France	M	Medium–late	Free	Low	Thin	–	Confectionary, marrons glacés
Dayouli	M	–	China	L–XL	Medium–late	–	–	–	–	Cooked
De Cedo	S	Famosa de Cedo	Spain	M	Medium	Free	Nul	Thin	M	Flour, candied, marmalade, purée
De Lemos	S	Lemás, Lemés, De Sarria	Spain	M	Late–very late	Free	High	Variable	M	Fresh, candied, flour, marmalade, purée
De Pata	S	Mansa	Spain	XS	Late	Free–Partial	Nul	Thin	L	Flour, marmalade, purée
De Sala	S	–	Spain	M	Late–very late	Free	Nul	Variable	L	Fresh
De San Miguel	S	Mingueira, Mingueiro	Spain	L	Late	Complete	Moderate	Very thin	M	Fresh, flour, marmalade, purée
Del Haya	S	–	Spain	XS	Late	Partial	Low	Very thin	L	Flour, marmalade, purée
Demanda	S	–	Portugal	M	Late	Free	Low	Thick	–	–
Dengorou	C	–	Japan	XL	Medium	Complete	Moderate	–	M	–

(Continued)

Cultivar	Species	Synonyms	Origin	Nut Size	Ripening Time	Pellicle Adhesion	% of Doubles	Nut Stipes	Catkins	Remarks/ Main Utilizations
Desgrañadiza	S	Desgrañadizas, Desgrañadoira	Spain	S	Late–very late	Partial	Nul	Thin	B	Flour, marmalade, purée
Dieguina	S	–	Spain	M	Medium–late	Complete	Moderate	Very thin	L	Fresh, candied
Do País	S	–	Spain	XS	Late–very late	Free	Nul	Thick	M	Flour, marmalade, purée
Donoso	S	–	Spain	S	Late–very late	Free–Partial	Low	Very thin	L	Flour, marmalade, purée
Dorée de Lyon	S	Marron de Lyon	France	M–L	Medium–late	Partial–Complete	Very high	Thin	B	Fresh
Doriga	S	Loriga, Origa	Spain	XS	Medium–late	Partial–Partial	High	Thin	L	Different genotypes/ Fresh, flour, marmalade, purée
Douglas hybrid	M × D	–	USA	L	Late	–	–	–	–	–
Duancibanli	M	–	China	M	early	Free	–	–	–	Cooked, roasted
Duanzha	M	–	China	L	Early– medium	Free	–	–	–	Cooked
Dunstan hybrid	M × D	–	USA	M	–	–	–	–	–	–
Dursun Kestanesi	S	–	Turkey	M	Medium	Partial	–	–	L	I(+)

(Continued)

Cultivar	Species	Synonyms	Origin	Nut Size	Ripening Time	Pellicle Adhesion	% of Doubles	Nut Stipes	Catkins	Remarks/ Main Utilizations
Eaton	$M \times C \times S$	–	USA	L	Early	–	–	–	–	–
Eaton River	$M \times C \times S$	–	USA	–	–	–	–	–	–	–
Enanishiki	C	–	Japan	L	Very early	Complete	–	–	M	–
Enxerta	S	–	Portugal	XS	–	Free	Nul	–	–	I(–)
Erfelek	S	–	Turkey	XS	Medium	Free	Low	Thick	L	I(–)
Erhuangzao	M	–	China	M–L	Early	Free	–	–	–	Cooked
Ersinop	S	–	Turkey	XS	Very early	Free	Low	Thick	L	I(–)
Erxinzao	M	–	China	M–L	Early–medium	Free	–	–	–	Cooked
Eryayla	S	–	Turkey	XS	Early	Free	Nul	Thin	L	I(–)
Escamplero	S	–	Spain	M	Late–very late	Partial	Nul	Thin–very thin	L	Fresh
Eurobella	$S \times P$	Silverleaf	USA	M–L	–	–	–	–	–	Pollinizer
Everfresh	M	–	USA	–	–	–	–	–	–	–
Evrytania	S	Karpenisi	Greece	M–L	Medium–late	Free	Low	–	–	Fresh, candied
Famosa	S	Marela, Amarela	Spain	M	Late–very late	Free	Nul	Thin	B	Fresh, flour, marmalade, purée
Fano	S	–	Spain	S	Medium–late	Free	Low	Very thin	L	Flour, marmalade, purée
Feltrona	S	Seronda, Pelona	Spain	XS	Medium–late	Free	Low	Thin	L	Flour, marmalade, purée

(Continued)

Cultivar	Species	Synonyms	Origin	Nut Size	Ripening Time	Pellicle Adhesion	% of Doubles	Nut Stipes	Catkins	Remarks/Main Utilizations
Ferosacre	C × S	–	France	–	–	–	–	–	–	Worldwide distributed/Rootstock
Firdola	S	–	Turkey	S	Very early	Partial	–	–	L	I(−)
Ford's Tall	M	–	USA	M	–	–	–	–	–	–
Formosa (Madeira)	S	–	Portugal	L	Late–very late	–	–	–	–	–
Frattona	S	–	Italy	M	Late	Free	Low	Thin	L	I(−)/Dried, flour
Frente Larga	S	–	Spain	M	–	Partial	Low	Thin	L	Fresh
Fukunami	C	–	Japan	XL	Early–medium	Complete	–	–	M	–
Fukunishi	C	–	Japan	L	Early–medium	Complete	Moderate	–	B	–
Gabiana	S	Gabbiana	Italy	XS	Late	Free	Low	Thin	L	I(−)/Dried, flour
Galega	S	Colunga, Zapatona, Galliciana	Spain	S–M	Medium–late	Partial	Moderate	Thin	L	Different genotypes/Fresh
Ganne	C	Shougatsu	Japan	XL	Medium–late	Complete	Low	–	–	–
Garfagnina	S	Scuna	Italy	M	Early–medium	Partial	–	–	–	Fresh
Garfagnina	S	Scuna	Italy	M	Early–medium	–	–	–	–	Fresh

(Continued)

Cultivar	Species	Synonyms	Origin	Nut Size	Ripening Time	Pellicle Adhesion	% of Doubles	Nut Stipes	Catkins	Remarks/ Main Utilizations
Garrida	S	Garrido	Spain	L	Late–very late	Free	Low	Thin	B	Fresh, flour, marmalade, purée, timber
Garriga	S	–	Spain	XS	Late	Partial	Nul	Thick	B	Flour, marmalade, purée
Garrone Nero	S	–	Italy	M	Medium	Partial	Low	Thin	A	I(–)/Fresh
Garrone Rosso	S	Caruna	Italy	L	Medium	Free	Low	Thick	A	I(–)/Fresh, marrons glacés
Gellatly # 1	M	–	USA	–	–	–	–	–	–	–
Genotte	S	–	Italy	S	Medium	Partial	–	–	–	Fresh
Gentile	S	–	Italy	S	Medium	Free	Moderate	Thin	L	I(+)/Fresh
Gideon	M	–	USA	M	Medium	–	–	–	–	–
Ginrei	C	Chuutan A	Japan	L	Early–medium	Complete	–	–	M	–
Ginyose	C	Tanba	Japan	XL	Early–medium	Complete	Moderate	–	M	–
Gioviasca	S	Giuvigliasca	Italy	M–L	Medium	Partial	Moderate	Thick	–	I(–)/Fresh
Gosha	C	–	Japan	M	Medium	Complete	Moderate	–	M	–
Grüa	S	De La Grua	Spain	M	Medium– very late	Partial	Moderate	Very thin	L	Fresh
Gureni	S	–	Romania	M	Early– medium	–	–	–	–	High productivity
Hacıbiş	S	–	Turkey	S	Very early	Partial	–	–	L	I(–)

(Continued)

Cultivar	Species	Synonyms	Origin	Nut Size	Ripening Time	Pellicle Adhesion	% of Doubles	Nut Stipes	Catkins	Remarks/ Main Utilizations
Haciömer	S	–	Turkey	M	Medium	Free	–	–	M	I(–)
Hakury	M	–	Japan					–	–	–
Hassaku	C		Japan	S	Very early	Complete	–	–	–	–
Hatayaooguri	C		Japan	XL	Medium	Complete	–	–	L	Very large nuts
Hayashi 1	C × M	–	Japan	–	–	Complete	–	–	–	–
Hayashi amaguri	C × M	–	Japan	–	–	Complete	–	–	–	–
Helechal	S	–	Spain	M	Medium–late	Partial	Moderate	Thin	B	Fresh
Higan	C	Inasabonguri	Japan	S	Early–medium	Complete	–	–	L	–
Hobita	S	–	Romania	M	Medium	–	–	–	–	High productivity
Hokugin	C	Toyogin	Japan	L	Very early	Complete	High	–	M	–
Hong Kong	M	–	USA	M		–		–	–	–
Hongguangyouli	M	–	China	S–M	Late	Free	–	–	–	Cooked
Hongli	M	–	China	M–L	Early	Free	Low	–	–	Roasted
Hongmaozao	M	–	China	L	Very early	–	–	–	–	Cooked
Horrón	S	Hórrea, Nizón	Spain	M	Late–very late	Partial	Low	Very thin	M	Fresh, flour, marmalade, purée
Houji 360	M	Miyazakishinaguri, Yunba 1 gou, Houji 23	Japan	S	Medium–late	Free	Low	–	B	–

(Continued)

Cultivar	Species	Synonyms	Origin	Nut Size	Ripening Time	Pellicle Adhesion	% of Doubles	Nut Stipes	Catkins	Remarks/ Main Utilizations
Huaguang	M	–	China	S	Very early–early	–	–	–	–	Cooked, roasted
Huaqiaobanli	M	–	China	L	Very early	–	–	–	–	Cooked
Hubeiyouli	M	–	China	S–M	Early–medium	Free	–	–	–	Cooked
Huihuangyouli	M	–	China	S–M	Late	Free	–	–	–	Cooked
Ibuki	C	–	Japan	XL	Very early–early	Complete	–	–	L	–
Ichiemon	C	–	Japan	L	Medium	Complete	Low	–	M	–
Ichikawawase	C	–	Japan	M	Early–medium	Complete	–	–	L	–
Idae	C × M	–	South Korea	XL	Early	Complete	–	–	–	–
Imakita	C	–	Japan	M	Medium	Complete	Low	–	M	–
Imperiale	S	–	France	M–L	Early	–	–	–	–	Pollinizer
Injerta (Asturias)	S	–	Spain	XS	Medium–late	Free	Very high	Very thin	M	Fresh, flour, marmalade, purée
Injerta (Bierzo)	S	Gallego	Spain	XS	Medium–late	Free	Moderate	Thin	M	Flour, marmalade, purée
Injerta (Extremadura)	S	–	Spain	L	Late	Free	Nul	Thin	L	Marron glacé
Insidina	S	–	France	M–L	Early	–	–	–	–	Fresh, processed

(Continued)

Cultivar	Species	Synonyms	Origin	Nut Size	Ripening Time	Pellicle Adhesion	% of Doubles	Nut Stipes	Catkins	Remarks/Main Utilizations
Inxerta	S	–	Spain	M	Late–very late	Partial	Low	Thick	M	Different genotypes/Fresh, flour, marmalade, purée
Ipharra	C	Iphara	France	L–XL	Early	–	–	–	–	I(–)/Fresh
Ishizuki	C	Ishizuchi	Japan	XL	Medium	Complete	Low	–	M	–
Iza	S	–	Romania	S	Early–medium	–	–	–	–	High productivity, precocity
Izumo	C	–	Japan	L	Early	Complete	–	–	–	–
Jabuda	S	–	Spain	L	Medium–very late	Partial	Low	Thin–very thin	L	Marron glacé
Jahong	C	–	South Korea	M	Medium	Free	Low	–	–	–
Jersey Gem	M	–	USA	S	–	–	–	–	–	–
Jiandingyouli	M	–	China	M	Early	Free	Low	–	–	Roasted
Jiaoci	M	–	China	S–M	Early–medium	Free	–	–	–	Cooked
Jiaozha	M	–	China	L	Early	Free	Low	–	–	Cooked
Jiebanli	M	–	China	M–L	Early	–	–	–	–	Cooked, roasted
Jiujiazhong	M	–	China	M–L	Early–medium	Free	–	–	–	Cooked

(Continued)

Cultivar	Species	Synonyms	Origin	Nut Size	Ripening Time	Pellicle Adhesion	% of Doubles	Nut Stipes	Catkins	Remarks/ Main Utilizations
Jiuyuehan	M	–	China	M	Late	Free	–	–	–	Cooked
Judia	S	–	Portugal	L	Late	Free	Nul	Thin	–	–
Kanotsume	C	–	Japan	L	Medium	Complete	Low	–	M	–
Karamehmet	S	–	Turkey	XS	Very early	Partial	–	–	L	I(–)
Kasaharawase	C	–	Japan	L	Medium	Complete	–	–	L	–
Katayama	C	–	Japan	XL	Medium–late	Complete	Low	–	L	–
Kenagaginyose	C	–	Japan	XL	Early– medium	Complete	Moderate	–	B	–
Kinseki	C	–	Japan	L	Medium	Complete	Low	–	M	–
Kinshuu	C	Kannabe	Japan	XL	Early– medium	Complete	Low	–	M	–
Kinyoshi	C	–	Japan	XL	Medium	Complete	Low	–	M	–
Kohr	H	–	USA	L	–	–	–	–	–	–
Konishiki	C	Gora	Japan	L	Medium–late	Complete	Low	–	M	–
Kretiko (Cretan)	S	–	Greece	M	Medium–late	Free	Nul	–	–	Fresh
Kuili	M	–	China	L	Early	–	–	–	–	Cooked

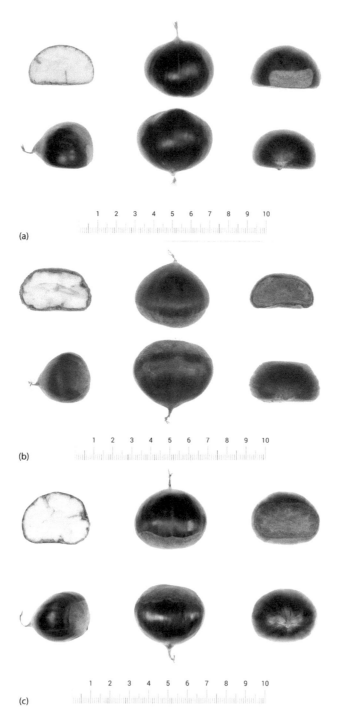

FIGURE 4.2   (a) Garrone rosso; (b) Ginyose; (c) Hakury.

(Continued)

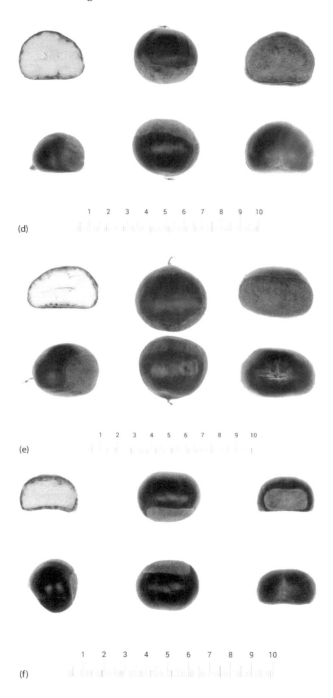

FIGURE 4.2 (Continued)   (d) Idae; (e) Ishizuki; and (f) Jersey Gem. (a, b, e,: Courtesy of Betta, M.; c: Courtesy of Gamba, G.; d, f: Courtesy of Riondato, I.)

Cultivar	Species	Synonyms	Origin	Nut Size	Ripening Time	Pellicle Adhesion	% of Doubles	Nut Stipes	Catkins	Remarks/Main Utilizations
Kunimi	C	–	Japan	XL	Very early–early	Complete	Low	–	–	–
Labor Day	C	–	USA	–	–	–	–	–	–	–
Lada	S	–	Portugal	M	Late	Free	Nul	Thin–thick	–	–
Laguépie	S	Marron de Laguépie, Rousse, Grosse Rousse	France	L	Medium–late	–	Very high	Thin	M	I(–)/Fresh
Lamela	S	–	Portugal	L	Late	Free	Nul	Thick	–	–
Largaña	S	Nargana, Parruca	Spain	XS	Medium–late	Free	Moderate	Thin–very thin	L	Flour, marmalade, purée
Leinova	S	Ley Nova	Spain	S	Medium–late	Partial	Moderate	Very thin	L	Flour, marmalade, purée
Lesvos	S	Agiasos	Greece	S	Early–medium	Free	Nul	–	–	Fresh, purée
Liaoli	M	–	China	L	Early	–	–	–	–	Cooked
Lindstrom#43	M	–	USA	M	–	–	–	–	–	–
Lisboa (Madeira)	S	–	Portugal	L	Medium	–	–	–	–	–
Lisio	S	Menudo, Liso	Spain	L	Medium–late	Partial	Low	Thin–very thin	L	Marron glacé
Liuyuebao	M	–	China	L	Early	Free	–	–	–	Cooked
Llanisca	S	Cofina	Spain	XS	Late	Free	High	Very thin	L	Flour, marmalade, purée

(Continued)

Cultivar	Species	Synonyms	Origin	Nut Size	Ripening Time	Pellicle Adhesion	% of Doubles	Nut Stipes	Catkins	Remarks/Main Utilizations
Longal	S	Enxerta	Portugal, Spain	M	Late–very late	Free	Low	Thin	B	Very big nuts
Loura	S	–	Spain	M–L	Late–very late	Partial	Nul–low	Thin	M	Different genotypes/Fresh, flour, marmalade, marron glacé, purée
Luccichente	S	Luccichente dell'Amiata	Italy	M–L	Medium	Partial	–	–	–	Fresh
Luguesa	S	Tarabelao	Spain	M	Late–very late	Partial	Low	Thin	M	Fresh, flour, marmalade, purée
Luina	S	–	Switzerland	S	Medium	Free	Low	–	A	Dried, flour
Lusenta	S × C	–	Italy	M	Early	Partial	Moderate	Thick	–	I(–)/Fresh
Luvall's Monster	Y	–	USA	L	–	–	–	–	–	–
Lyeroka	M × S	–	USA	M	Early	–	–	–	A	–
Maceirá	S	Mazaiño, Mazaiña	Spain	S	Late	Free	Low	Very thin	M	Fresh, flour, marmalade, purée
Macho	S	–	Spain	XL	–	Free	Low	Very thin	L	Marron glacé
Macit 55	Y	–	Turkey	L	Very early	Free	High	Very thin	L	On registration process/I(–)

(Continued)

Cultivar	Species	Synonyms	Origin	Nut Size	Ripening Time	Pellicle Adhesion	% of Doubles	Nut Stipes	Catkins	Remarks/Main Utilizations
Madonna	S	Canalina, di Canale, Castagna della Madonna	Italy	M–L	Very early	Partial	Very high	Thin	L	I(−)/Fresh, fruit
Mahmutmolla	S	−	Turkey	M	Early	Free	−	−	B	I(−)
Mansinha (Madeira)	S	−	Portugal	S	Medium	−	−	−	−	−
Manso	S	−	Spain	M	Late–very late	Free–Partial	Very high	−	M	Fresh
Maobanhong	M	−	China	L	Medium	−	−	−	−	Cooked
Mara	S	−	Romania	L	Early	−	−	−	−	Early bearing
Maraval	C × S	−	France	L–XL	Medium	Partial–Complete	Low	Thick	L	Worldwide distributed/Rootstock, I(−)
Marela	S	−	Spain	XS–S	Medium–late	Free	Very high	Thin	B	Different genotypes/Flour, marmalade, purée
Mariana	S	−	Spain	S	Medium–late	Partial	Moderate	Thin–very thin	L	Flour, marmalade, purée
Maridonne	S × C	−	France	L–XL	Late–very late	Partial	Low–moderate	Thick	M	Worldwide distributed/I(−)/Fresh, processed
Marigoule	C × S	−	France	L–XL	Early–medium	Free–Partial	Low	Thin	L	Worldwide distributed/I(−)/Fresh, processed

(Continued)

Cultivar	Species	Synonyms	Origin	Nut Size	Ripening Time	Pellicle Adhesion	% of Doubles	Nut Stipes	Catkins	Remarks/Main Utilizations
Marina	S	–	Spain	S	Late	Free	Very high	Thin–very thin	A	Flour, marmalade, purée
Marissard	C × S	–	France	L–XL	Medium	Free	High	Thick	–	Worldwide distributed/I(−)/Fresh
Marki	C	–	France	L	Early–medium	Partial	–	–	–	–
Marlhac	S × C	–	France	L–XL	Medium	Partial	Low–moderate	Thin	A	Worldwide distributed/Rootstock, I(−)/Fresh, cream
Marron Buono (di Marradi)	S	Marron buono, Marrone di Marradi	Italy	L	Medium	Free	Low	Thick	A	I(−)/Marrons glacés
Marron d'Olargues	S	Marron de Saint-Vincent	France	M–L	Medium–late	Free	Low–moderate	Thick	–	I(−)/Confectionary, marrons glacés
Marron de Chevanceaux	S		France	M–L	Medium–late	Partial	Nul	Thick	L	I(−)/Marrons glacés
Marron de Goujounac	S	–	France	L–XL	Medium	Partial	–	–	L	Pollinizer, I(−)/Fresh
Marron de Redon	S	Marron de la Saint-Jean	France	M–L	Medium	Partial	Low–moderate	Thin	–	I(−)

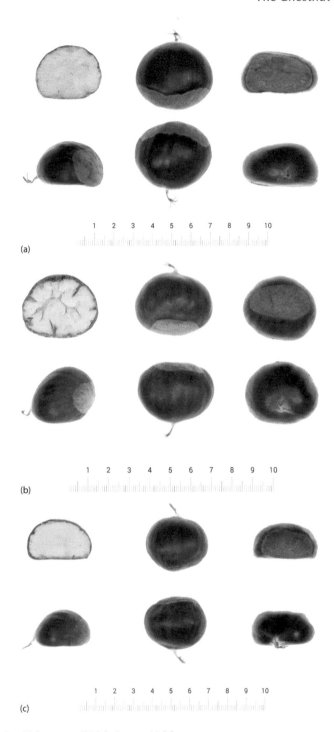

FIGURE 4.3   (a) Lusenta; (b) Madonna; (c) Mansa.

(Continued)

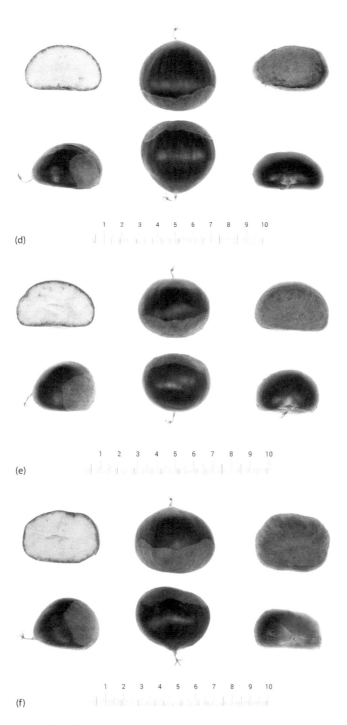

FIGURE 4.3 (Continued) (d) Maridonne; (e) Marigoule; and (f) Marissard. (a–f: Courtesy of Betta, M.)

Cultivar	Species	Synonyms	Origin	Nut Size	Ripening Time	Pellicle Adhesion	% of Doubles	Nut Stipes	Catkins	Remarks/Main Utilizations
Marron du Var	S	Marrouge, Marron du Luc, Marrougio, Marron de Collobrières	France	M–L	Late	Free	Moderate	Thick	–	I(–)/Confectionary, marrons glacés
Marrone Badia Coltibuono	S	–	Italy	L	Medium	Free	–	–	A	I(–)/Fresh
Marrone Borra Montesevero	S	–	Italy	L	Medium	Free	–	–	A	I(–)/Fresh
Marrone De Coppi	S	Marrone di Val Susa	Australia	L	Medium	Free	Low	Thick	A	I(–)/Fresh, marrons glacés, processed
Marrone dell'Isola d'Elba	S	–	Italy	M	Medium–late	Free	–	Thick	A	I(–)/Fresh
Marrone dell'Amiata	S	Marrone del Monte Amiata	Italy	M	Medium–late	Free	–	–	A	I(–)/Fresh, marrons glacés, processed
Marrone di Avellino	S	Marrone Avellinese, Santomango, Santimango	Italy	L	Early	Free	–	–	A	I(–)/Fresh, marrons glacés, processed
Marrone di Bosco Vittoria	S	–	Italy	M	Late	Free	–	–	A	I(–)/Fresh, marrons glacés, processed
Marrone di Caprarola	S	–	Italy	L	–	Free	–	–	A	I(–)/Fresh, marrons glacés, processed
Marrone di Caprese Michelangelo	S	–	Italy	L	Medium	Free	Low	Thick	A	I(–)/Fresh, marrons glacés, processed

(Continued)

Cultivar	Species	Synonyms	Origin	Nut Size	Ripening Time	Pellicle Adhesion	% of Doubles	Nut Stipes	Catkins	Remarks/Main Utilizations
Marrone di Castel Borello	S	–	Italy	L	Medium	Free	Low	Thick	A	I(–)/Fresh, marrons glacés, processed
Marrone di Castel del Rio	S	–	Italy	L	Early–medium	Free	Low	Thick	A	I(–)/Fresh, marrons glacés, processed
Marrone di Castiglione	S	–	Italy	–	–	Free	–	–	A	I(–)/Fresh, marrons glacés, processed
Marrone di Castiglione dei Pepoli	S	–	Italy	L	Medium	Free	Low	Thick	A	I(–)/Fresh, marrons glacés, processed
Marrone di Chiusa Pesio	S	–	Italy	L	Medium	Free	Nul–low	Thick	A	I(–)/Fresh, marrons glacés, processed
Marrone di Citta' di Castello	S	–	Italy	L	Medium	Free	Low	Thick	A	I(–)/Fresh, marrons glacés, processed
Marrone di Combai	S	–	Italy	M	Medium	Free	–	–	A	I(–)/Fresh, marrons glacés, processed
Marrone di Forli	S	–	Italy	L	Medium–late	Free	–	–	A	I(–)/Fresh, marrons glacés, processed
Marrone di Gavignano	S	–	Italy	L	Medium–late	Free	Low	Thick	A	I(–)/Fresh, marrons glacés, processed
Marrone di Luserna	S	Savatua	Italy	M	Late	Free	Nul	Thick	A	I(–)/Fresh, marrons glacés, processed
Marrone di Monfenera	S	–	Italy	M	Medium	Free	–	–	A	I(–)/Fresh, marrons glacés, processed
Marrone di Montemarano	S	–	Italy	L	Medium	Free	Low	–	A	I(–)/Fresh, marrons glacés, processed

(*Continued*)

Cultivar	Species	Synonyms	Origin	Nut Size	Ripening Time	Pellicle Adhesion	% of Doubles	Nut Stipes	Catkins	Remarks/Main Utilizations
Marrone di Montepastore	S	–	Italy	M	Early	Free	–	–	A	I(–)/Fresh, marrons glacés, processed
Marrone di Palazzo del Pero	S	–	Italy	–	–	Free	–	–	A	I(–)/Fresh, marrons glacés, processed
Marrone di Roccadaspide	S	–	Italy	M–L	Early–medium	Free	Low	–	A	I(–)/Fresh, marrons glacés, processed
Marrone di Roccaverano	S	–	Italy	M	Medium	Free	Low	–	A	I(–)/Fresh, marrons glacés, processed
Marrone di San Mauro	S	–	Italy	M	Medium	Free	–	–	A	I(–)/Fresh, marrons glacés, processed
Marrone di Segusino	S	–	Italy	L	–	Free	–	–	A	I(–)/Fresh, marrons glacés, processed
Marrone di Serino	S	–	Italy	–	–	Free	–	–	A	I(–)/Fresh, marrons glacés, processed
Marrone di Stia	S	–	Italy	M	Medium	Free	–	–	A	I(–)/Fresh, marrons glacés, processed
Marrone di Val Susa	S	Marrone di San Giorio	Italy	L	Medium	Free	Nul–low	Thick	A	I(–)/Fresh, marrons glacés, processed
Marrone di Villar Pellice	S	–	Italy	L	Late	Free	Nul	Thick	A	I(–)/Fresh, marrons glacés, processed
Marrone di Viterbo	S	–	Italy	L	Medium	Free	–	–	A	I(–)/Fresh, marrons glacés, processed
Marrone Fiorentino	S	–	Italy	L	Early–medium	Free	Low	Thick	A	I(–)/Fresh, marrons glacés, processed
Marrubia	S	–	Italy	M	Medium–late	Free	Very high	Thick	A	I(–)/Fresh

(Continued)

Cultivar	Species	Synonyms	Origin	Nut Size	Ripening Time	Pellicle Adhesion	% of Doubles	Nut Stipes	Catkins	Remarks/Main Utilizations
Marsol	C × S	–	France	L–XL	Early–medium	Partial	Low	Thick	L	Worldwide distributed/Rootstock, I(–)
Martaínha	S	–	Portugal	L	Medium	Free	Moderate	Thick	–	–
Marzapanara	S	–	Italy	L	–	Partial	–	–	L	Fresh
Matabei	C	–	Japan	XL	Medium	Complete	Moderate	–	M	–
Matancero	S	–	Spain	XS	Late	Partial	Low	Very thin	M	Flour, marmalade, purée
Mediana	S	Injerta, Palaciana, Parede, Pelgo, Ximara	Spain	XS	Medium–late	Partial	Very high	Thin–very thin	L	Flour, marmalade, purée
Meiling	M	–	USA	S	–	–	–	–	–	–
Mercogliana	S	Merculiana	Italy	M	Early–medium	Partial	–	–	L	Fresh
Merle	S	–	France	M–L	Medium	Complete	Moderate	Thick	–	Rootstock/Fresh
Miguelina	S	–	Spain	S	Medium–late	Free	Very high	Thin	A	Different genotypes/Flour, marmalade, purée
Mikuri	C	–	Japan	XL	Medium	Complete	Low	–	M	–
Miller 72-76	M	–	USA	–	–	–	–	–	–	–
Mipung	C	–	South Korea	XL	Late	–	–	–	–	–
Misericordia	S	–	Portugal	XS	Medium	Free	Nul	Thin	–	–
Mollar	S	–	Spain	XS	Medium–late	Free	High	Very thin	L	Flour, marmalade, purée

(*Continued*)

Cultivar	Species	Synonyms	Origin	Nut Size	Ripening Time	Pellicle Adhesion	% of Doubles	Nut Stipes	Catkins	Remarks/Main Utilizations
Monfortina	S	Arial	Spain	XS	Late–very late	Complete	Nul	Thick	L	Pollinizer/Flour, marmalade, purée
Montagne	S	Grosse de Loubejac, Marron de Villefranche	France	M	Medium–late	Partial	Moderate–high	Thick	L	Fresh, processed
Montemarano	S	Montellese, Palummina	Italy	L–XL	Medium–late	Partial	–	Thick	–	Fresh, dried
Montesín	S	Montesín de la Cruz	Spain	S	Medium–late	–	Very high	Very thin	B	Flour, marmalade, purée
Moreira	S		Portugal	XS	Medium	Free	Moderate	–	–	–
Moriwase	C	–	Japan	L	Very early	Complete	–	–	–	–
Mossbarger	M	–	USA	L	–	Free	–	–	–	–
Mourisca	S	Mourisco	Spain	XS	Medium–late	Free	High	Very thin	–	Flour, marmalade, purée
Mulata	S	Picoclaro	Spain	S	Late–very late	Complete	Moderate	Thin–very thin	M	Flour, marmalade, purée
Mulata (Azores)	S	–	Portugal	S	Early	–	–	–	–	–
Myoka	M × S	–	USA	M	–	Free	–	–	–	–
Nakatetanba	C	Chuuwase, Hayadama, Wasechoubei	Japan	XL	Very early–early	Complete	–	–	L	–
Nanking	M	–	USA	M–L	Late	–	–	–	–	–
Napoletana	S	Riccia napoletana	Italy	M	Early	Partial	–	–	L	Fresh
Navexa	S	Naveixa	Spain	M	Medium–late	Partial	High	Thin–very thin	L	Fresh, flour, marmalade, purée
Negra	S	–	Portugal	M	Late	Free	Low	–	–	–

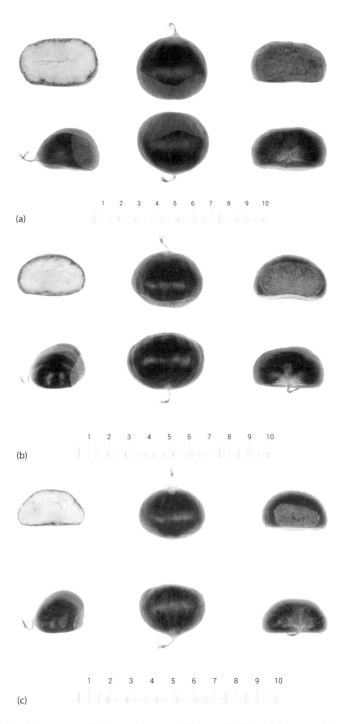

FIGURE 4.4 (a) Marlhac; (b) Marron Buono (di Marradi IGP); (c) Marrone Fiorentino.

(*Continued*)

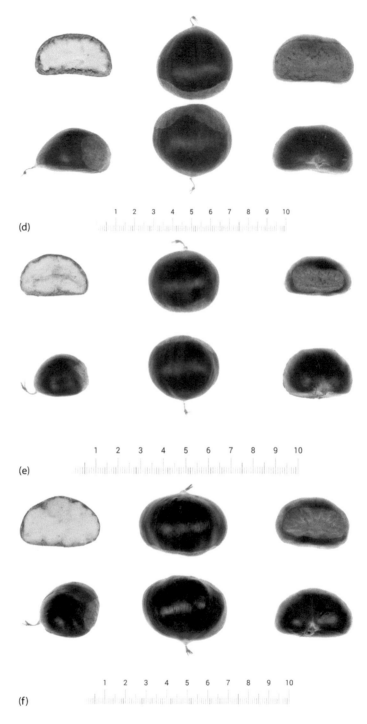

FIGURE 4.4 (Continued)   (d) Marsol; (e) Merle; and (f) Montagne. (a–f: Courtesy of Betta, M.)

Cultivar	Species	Synonyms	Origin	Nut Size	Ripening Time	Pellicle Adhesion	% of Doubles	Nut Stipes	Catkins	Remarks/Main Utilizations
Negral	S	Blanca, Courelá, Médulas, Riá, Vaquera, Vileta, Xabrega	Spain	S–M	Medium–late	Free–Partial	Variable	Thin–very thin	L	Different genotypes/Fresh
Negral	S	–	Portugal	L	Late	Free	Low	Thin–thick	–	Pollinizer/Fresh, marmalade, purée
Negrera	S	Negral	Spain	XS	Medium–late	Free	Moderate	Very thin	L	Flour, marmalade, purée
Negro	S	–	Spain	L	–	Complete	Very high	Very thin	L	Fresh
Neirana	S	Neirane	Italy	S	Medium	Partial	Very high	Thin	–	I(–)/Fresh
Nevada	S × C	–	USA	M–L	Late	–	–	–	–	Pollinizer
Ninomiya	C	–	Japan	XL	Medium	Complete	–	–	–	–
Nzierta	S	Inserta, Nzerta	Italy	M	Medium–late	Partial	–	–	L	Fresh
Obiacco	S	–	Italy	M	Medium–late	Free	–	–	–	I(–)/Fresh, processed
Obiwase	C	–	Japan	S	Very early	Complete	–	–	–	–
Obuse 2	C	–	Japan	M	Medium	Complete	Low	–	M	–
Odai	C	–	Japan	XL	Very early–early	Complete	–	–	L	–
Ogawateteuchi	C	–	Japan	L	Very early–early	Complete	Moderate	–	L	–
Okei	H	–	USA	XL	–	Partial	–	–	–	–
Okkwang	C	–	South Korea	L	Early	–	–	–	–	–
Ooharaguri	C	–	Japan	M	Very early–early	Complete	–	–	M	–
Ookoma	C	–	Japan	–	–	Complete	–	–	–	–

(Continued)

Cultivar	Species	Synonyms	Origin	Nut Size	Ripening Time	Pellicle Adhesion	% of Doubles	Nut Stipes	Catkins	Remarks/Main Utilizations
Ordaliega	S	–	Spain	S	Medium–late	Free	Low	Very thin	–	Flour, marmalade, purée
Orrin	M	–	USA	XS	Medium	–	–	–	–	–
Osaya	C	–	Japan	M	Very early–early	Complete	–	–	M	–
Osmanoğlu	S	–	Turkey	M	Early	Partial	Nul	Thick	A	I(+)
Otomune	C	–	Japan	L	Early–medium	Complete	–	–	M	–
Ouriza	S	Ouriceira	Spain	XS	Late–very late	Free	Nul	–	B	Flour, marmalade, purée
Outeira	S	Outeiro	Spain	XS	Late–very late	Free	Nul	Thin	B	Flour, marmalade, purée, timber
Ozark chinkapin	P	–	USA	XS	Very early	–	–	–	–	C. pumila var. ozarkensis/ Ornamental, timber
Pagana	S	–	Spain	XS	Medium	Partial	Very high	Thin–very thin	B	Flour, marmalade, purée
Palaciana	S	Injerta, Mediana, Parede, Pelgo, Ximara	Spain	S	Medium–late	Partial	Nul	Thin	L	Flour, marmalade, purée
Pallaregas	S	Pallarega, Pallaregos, Loura	Spain	XS	Late–very late	Partial	Nul	Very thin	M	Flour, marmalade, purée
Palummella	S	Palummina	Italy	L	Medium–late	Partial	–	–	–	Fresh, processed
Palummina	S	–	Italy	L	Medium–late	Partial	–	–	–	Fresh, processed

(Continued)

Cultivar	Species	Synonyms	Origin	Nut Size	Ripening Time	Pellicle Adhesion	% of Doubles	Nut Stipes	Catkins	Remarks/Main Utilizations
Panchina	*S*	–	Spain	S	Medium–late	Free	High	Thin–very thin	L	Flour, marmalade, purée
Paradesa	*S*	Parede	Spain	M	Late–very late	Free	Nul	–	M	Flour, marmalade, purée
Paragon	*S × D*	–	USA	M	Medium	–	–	–	–	–
Parede	*S*	Padana, Paradesa, Xidra	Spain	XS	Late–very late	Free	Low–moderate	Very thin	M	Different genotypes/ Fresh, marmalade, purée, timber
Passa	*S*	–	Portugal	L	Medium	Free	Low	–	–	–
Pastinese	*S*	–	Italy	S	Late	Partial	–	–	–	Flour
Pastorese	*S*	–	Italy	S	Late	Partial	–	–	–	Flour
Patacuda	*S*	–	Spain	L	Late–very late	Variable	Low	Thin	M	Flour, marmalade, purée
Payne	*M*	–	USA	L	–	–	–	–	–	–
Peach	*M*	–	USA	M	Medium	–	–	–	–	–
Pelada	*S*	Pelado	Spain	M	Late–very late	Free	Moderate	Thin	M	Flour, marmalade, purée
Pelgo	*S*	Xifmara	Spain	XS	Medium–late	Free	High	Thin	L	Flour, marmalade, purée
Pellegrine	*S*	–	France	–	Medium	Partial	–	–	L	Fresh, marmalade
Pelón	*S*	Mollar	Spain	M	Late–very late	Free–Partial	Low	Thin–very thin	L	Fresh
Pelona	*S*	Feltrona, Seronda	Spain	S–L	Medium–late	Free	Low–moderate	Thin	L	Different genotypes/ Flour, marmalade, purée

(Continued)

Cultivar	Species	Synonyms	Origin	Nut Size	Ripening Time	Pellicle Adhesion	% of Doubles	Nut Stipes	Catkins	Remarks/Main Utilizations
Peloño	S	–	Spain	S	Medium–late	Free–Partial	Nul	Thin–very thin	A	Fresh, flour, marmalade, purée
Pelosa	S	–	Italy	XS	Medium	Partial	–	–	A	Dried, flour
Pelosa	S	Peluses	Spain	M	Late	Partial	Moderate	Thin–very thin	L	Fresh
Peluda	S	–	Spain	M	Late–very late	Free	Nul	Thin	M	Fresh, flour, marmalade, purée
Peludo	S	–	Spain	XS	Late	Free	High	Thin	L	Flour, marmalade, purée
Perry	M	Lindstrom #93	USA	S	–	–	–	–	–	–
Picoclaro	S	–	Spain	XS	Late–very late	Partial–Complete	Moderate	Thin–very thin	L	Flour, marmalade, purée
Picona	S	–	Spain	S	Late–very late	Free	Very high	Thick	M	Flour, marmalade, purée
Picuda	S	–	Spain	S–M	Late–very late	Free–Partial	Null–low	Thin–very thin	L	Fresh, flour, marmalade, purée
Pilonga	S	–	Spain	L	Early–medium	Complete	Moderate	Thin–very thin	L	Different genotypes/Fresh
Piñero	S	–	Spain	S	Very late	Partial	Nul	Thin–very thin	B	Flour, marmalade, purée
Pistolese	S	–	Italy	M	Early	Partial	–	–	–	Fresh, flour
Planta Alajar	S	–	Spain	L	Medium–late	Complete	High	Thin–very thin	M	Fresh
Polegre	S	–	Spain	M	Late	Partial	Low	Thin	L	Fresh
Politora	S	–	Italy	M	Medium–late	Partial	–	–	–	Fruit and timber production

(Continued)

Cultivar	Species	Synonyms	Origin	Nut Size	Ripening Time	Pellicle Adhesion	% of Doubles	Nut Stipes	Catkins	Remarks/Main Utilizations
Polovragi	S	–	Romania	XL	Very late	–	–	–	–	Fresh
Pontecosa	S	Pontecose, Pontecosi, Puntecosa, Punticosa, Punticoso	Italy	M	Medium–late	Partial	–	–	–	Fresh, flour
Porosuke	C	–	Japan	L	Very early	Free	Low	–	L	–
Porotan	C	–	Japan	XL	Very early–early	Free	Low	–	L	–
Porteliña	S	Porteliñas	Spain	M	Late–very late	Free	Low	Thick	M	Candied, flour, marmalade, purée
Portuguesa	S	Portugués	Spain	XS	Late–very late	Free	Very high	Very thin	A	Flour, marmalade, purée
Pozoredondo	S	Pozo Redondo, De Pozo Redondo	Spain	S	Late–very late	Free	Low	Thin	M	Fresh, flour, marmalade, purée
Praga d'Afora	S	Puga d'Afora	Spain	M	Late–very late	Free	Low	Thick	B	Fresh, flour, marmalade, purée
Praga do Bolo	S	Puga do Bolo	Spain	M	Late–very late	Partial	Nul	Thin	B	Different genotypes/ Fresh, flour, marmalade, purée
Précoce Migoule	C × S	–	France	L	Early	Partial	Very high	Thin	L	Worldwide distributed/I(−)/ Fresh

(Continued)

Cultivar	Species	Synonyms	Origin	Nut Size	Ripening Time	Pellicle Adhesion	% of Doubles	Nut Stipes	Catkins	Remarks/Main Utilizations
Précoce Ronde des Vans	S	Preminche, Sardounencho, Vivaraise	France	S	Very early–early	Partial	Very high	Thick	–	I(–)/Fresh
Presa	S	–	Spain	M	Late–very late	Partial	Low	Thick	M	Fresh, marmalade, purée, timber
Preta (Madeira)	S	–	Portugal	XS	Late	–	–	–	–	–
Prigorie	S	–	Romania	M	Very late	–	–	–	–	High productivity
Primato	C × S	–	Italy	L	Very early	Free	Low	Thick	–	I(–)/Fresh
Primemura	S	–	Italy	M	Medium	Partial	–	Thick	–	Fresh
Primitiva di Roccamonfina	S	–	Italy	M	Early	Partial	–	–	L	I(–)/Fresh
Primitiva Riccia	S	–	Italy	M	Early	Partial	–	–	L	Fresh
Prolific	S	–	USA	–	–	–	–	–	–	–
Puga	S	De Puga	Spain	XS	Late–very late	Free	Low	Thick	M	Flour, marmalade, purée
Puga do Receiro	S	–	Spain	XS	Late–very late	Free	Low	Thick	B	Flour, marmalade, purée
Purton's Pride	–	–	Australia	–	–	–	–	–	–	–
Qiancidabanli	M	–	China	XL	Early	Free	Low	–	–	Cooked, roasted
Qing	M	–	USA	M–L	–	Free	–	–	–	–
Qingmaoruanci	M	–	China	M–L	Medium	–	–	–	–	Cooked, roasted
Qingmaozao	M	–	China	L	Early	–	–	–	–	Cooked, roasted
Qingzha	M	–	China	L	Early–medium	Free	–	–	–	Cooked, roasted

(Continued)

Cultivar	Species	Synonyms	Origin	Nut Size	Ripening Time	Pellicle Adhesion	% of Doubles	Nut Stipes	Catkins	Remarks/Main Utilizations
Queshanziyouli	M	–	China	L	Early	–	–	–	–	Cooked
Raigona	S	–	Spain	XS	Late-very late	Free	Moderate-high	Thin	B	Different genotypes/Flour, marmalade, purée
Ramiega	S	Ramuniega	Spain	S	Medium-late	Free	Low	Thin-very thin	L	Flour, marmalade, purée
Ranaz	S	–	Italy	M	Early-medium	Partial	–	–	–	Fresh
Rañuda	S	Ramuda	Spain	–	Late	–	–	–	B	–
Rapada Galicia	S	Rapado	Spain	XS–S	Late	Free	Low-moderate	Thick	M	Fresh, marmalade, purée, timber
Rapega	S	Verdello	Spain	S	Medium-late	Free	Very high	Thin	A	Flour, marmalade, purée
Rapuca	S	–	Spain	XS	Medium-late	Partial	Low-moderate	Very thin	M	Different genotypes/Flour, marmalade, purée
Ravexa	S	–	Spain	XS	Medium	Free	Moderate	Thin	L	Flour, marmalade, purée
Reborda	S	–	Portugal	M–L	Late	Free	Low	–	–	–
Red Spanish	–	–	Australia	–	–	–	–	–	–	–
Redonda	S	–	Portugal	S	Medium	Free	Nul	–	–	–
Redondo	S	Redonda	Spain	S–M	Late-very late	Partial	Low	Thick	L	Different genotypes/Pollinizer/Flour, marmalade, purée
Revival	M × D	–	USA	XS	–	–	–	–	–	–

(Continued)

Cultivar	Species	Synonyms	Origin	Nut Size	Ripening Time	Pellicle Adhesion	% of Doubles	Nut Stipes	Catkins	Remarks/Main Utilizations
Riá	S	Rial, Negral	Spain	S	Late	Complete	Low	Thin–very thin	L	Pollinizer/Fresh, flour, marmalade, purée
Ribeirá	S	Ribeirás, Ribeirao	Spain	M	Late–very late	Complete	Low	Very thin	M	Flour, marmalade, purée
Riggiola	S	Ricciola, Riggiuolo	Italy	M	Early–medium	Partial	–	–	–	Fresh
Riheiguri	C × M	Tanabeguri, Rihei	Japan	XL	Early–medium	Complete	Low	–	M	–
Rosenda	S	–	Spain	–	–	–	–	–	L	Pollinizer
Rossastra	S	–	Italy	XS	Late	Partial	–	–	–	Fresh, dried
Rossitta	S	–	Italy	S	Medium	–	–	–	–	Fresh
Rouffinette	S	Ruffinette	Italy	M	Medium	Free	–	Thin	–	Cooked
Rousse de Nay	S	Rouge de Nay, Grosse Rouge, Marron des Pyrénées	France	M–L	Medium	Partial	Very high	Thin	–	–
Roussette de Montpazier	S	–	France	–	–	Partial	–	–	–	–
Roxa	S	–	Spain	XS	Medium–late	Free	High	Thin	L	Flour, marmalade, purée
Roxina	S	–	Spain	S	Medium–late	Free	Low	Thin	M	Flour, marmalade, purée
Rozada	S	–	Spain	S	Late–very late	Partial	Nul	Very thin	M	Fresh, flour, marmalade, purée

(Continued)

Cultivar	Species	Synonyms	Origin	Nut Size	Ripening Time	Pellicle Adhesion	% of Doubles	Nut Stipes	Catkins	Remarks/Main Utilizations
Rubia	S	Rubiales	Spain	XS	Medium	Partial	Moderate	Very thin	L	Flour, marmalade, purée
Rubia Tardía	S	–	Spain	XL	Medium	Complete	High	Thick	L	Fresh
S. Martinho (Azores)	S	–	Portugal	M	Late	–	–	–	–	–
Saimyouji 1	C	–	Japan	XL	Early–medium	Complete	High	–	M	–
Salnesa	S	–	Spain	XS	Late	Free	Nul	Thin	L	Flour, marmalade, purée
San-Dae	C	–	South Korea	XL	Early	Complete	–	–	–	Confectionary
Santo antónio (Madeira)	S	–	Portugal	L	Late–very late	–	–	–	–	–
Sardonne	S	Sardoune, Marron de Villefort	France	L	Medium	Partial	High	Thick	A	Confectionary, marrons glacés
Sariaşlama	S	–	Turkey	L	Medium	Partial	–	–	L	I(+)
Sarvaschina	S	Selvaschina	Italy	S	Early	Partial	Moderate	Thick	L	Fresh

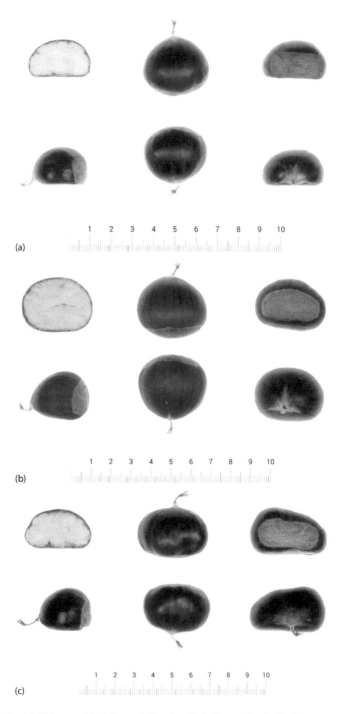

FIGURE 4.5   (a) Neirana; (b) Précoce Migoule; (c) Précoce Ronde des Vans.

(Continued)

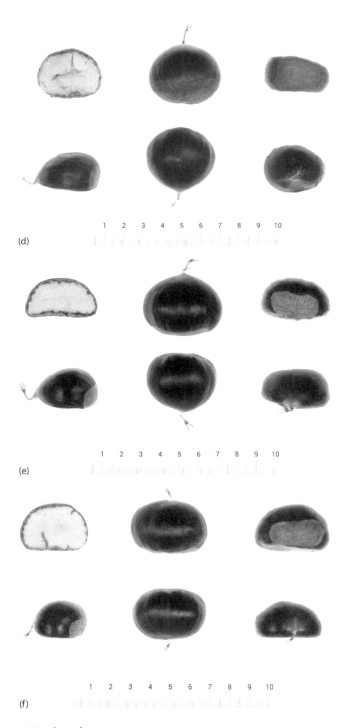

FIGURE 4.5 (Continued) (d) Primato; (e) Primemura; and (f) Rouffinette. (a, c, e, f: Courtesy of Betta, M.; b, d: Courtesy of Gamba, G.)

Cultivar	Species	Synonyms	Origin	Nut Size	Ripening Time	Pellicle Adhesion	% of Doubles	Nut Stipes	Catkins	Remarks/Main Utilizations
Savoye	S	—	France	M	Medium	Partial	—	—	—	Fresh
Seguin's chestnut	Se	Dwarf Chinese chestnut	China	XS	Very early	—	—	—	—	Firewood, ornamental
Senri	C	—	Japan	XL	Medium	Complete	—	—	L	—
Serdar	S	—	Turkey	XS	Late	Free	Nul	Thin	L	I(−)
Sergude	S	—	Spain	M	Late–very late	Partial	Low	Thin	L	Pollinizer/Fresh, flour, marmalade, purée
Serodia	S	Serodias, Ouriza, Serodia Pozo Mar, Monfortina, Verdello	Spain	M	Late–very late	Free	Low	Thick	L	Pollinizer/Fresh, flour, marmalade, purée
Sevillana	S	—	Spain	S	Medium–late	Partial	Very high	Thin	L	Flour, marmalade, purée
Seyrekdiken	S	—	Turkey	L	Medium	Partial	—	—	L	I(+)
Shangdonglaixidayouli	M	—	China	L	Early–medium	Free	—	—	—	Cooked
Shaoguan	M	—	China	M–L	Very early	—	—	—	—	Cooked
Shaoyangtali	M	—	China	M–L	Early	—	—	—	—	Cooked, roasted
Shencidabanli	M	—	China	L–XL	Medium–late	—	—	—	—	Cooked
Shichifukuwase	C	—	Japan	L	Very early–early	Complete	—	—	M	—
Shifeng	M	—	China	M–L	Early	—	—	—	—	Cooked, roasted

(Continued)

Cultivar	Species	Synonyms	Origin	Nut Size	Ripening Time	Pellicle Adhesion	% of Doubles	Nut Stipes	Catkins	Remarks/Main Utilizations
Shihou	C	Shiho	Japan	XL	Early–medium	Complete	Low	–	–	–
Shimokatsugi	C	–	Japan	XL	Medium–late	Complete	Low	–	M	–
Shing	M	–	USA	S	–	–	–	–	–	–
Shuhou	C	–	Japan	XL	Early–medium	Complete	–	–	–	–
Shuuhouwase	C	–	Japan	L	Very early–early	Complete	–	–	L	–
Siete Pernadas	S	–	Spain	S		Free–Partial	Moderate	Thin	B	Flour, marmalade, purée
Silverleaf	S	–	USA	L	–	Free	–	–	–	–
Simpson	M × S	–	USA	M	–	–	–	–	–	–
Siria	S	Ciria	Italy	XS	Late	Free	Low	Thick	L	I(–)/Dried, flour
Skioka	M × S	–	USA	M–L	Very late	–	–	–	–	Pollinizer
Skookum	M × S	–	USA	M	–	–	–	–	A	–
Sleeping Giant	M × D × C	–	USA	L	Early	Free	–	–	L	Fresh
Solenga	S	Solenca	Italy	M	Early	Free	–	–	L	Cooked, roasted
Songjiazao	M	–	China	S–M	Early	–	–	–	–	
Soulage Premiere	S	–	France	S–M	Early	Free–Partial	Nul–low	Thick	–	–
Sousa	S	–	Portugal	M	Late	Free	Low	Thin–thick	–	–
Soutogrande	S	–	Spain	L	Late–very late	Free	Nul	Thin	M	Fresh, flour, marmalade, purée
Spinalunga	S	–	Italy	XS	Late	Partial	Low	Thick	L	I(+)/Fresh

(Continued)

Cultivar	Species	Synonyms	Origin	Nut Size	Ripening Time	Pellicle Adhesion	% of Doubles	Nut Stipes	Catkins	Remarks/Main Utilizations
Szego	$C \times M \times P$	–	USA	–	–	–	–	–	–	–
Tablón	S	–	Spain	XS	Very late	Free	Low	Thin–very thin	M	Flour, marmalade, purée
Taishouwase	C	–	Japan	XL	Very early	Complete	–	–	–	–
Tajiriginyose	C	–	Japan	XL	Medium	Complete	Low	–	L	–
Tamanishiki	C	Ishizuka	Japan	XL	Very early–early	Complete	–	–	M	–
Tamatsukuri	C	–	Japan	–	–	–	–	–	–	–
Tamón	S	De Tamón	Spain	M	Late	Complete	Moderate	Very thin	L	Fresh
Tanoue 1 gou	C	–	Japan	XL	Early–medium	Complete	Low	–	L	–
Tanqiaobanli	M	–	China	L	Medium	–	–	–	–	Cooked
Tanzawa	C	–	Japan	XL	Very early	Complete	Low	–	L	Worldwide distributed
Tarabelao	S	Luguesa	Spain	M	Late	Partial–Complete	Nul	Thin	M	Fresh, flour, marmalade, purée
Tarvisò	S		Italy	M	Medium	Partial	–	–	M	Fresh

(Continued)

Cultivar	Species	Synonyms	Origin	Nut Size	Ripening Time	Pellicle Adhesion	% of Doubles	Nut Stipes	Catkins	Remarks/Main Utilizations
Temperá	S	Temperán, Temporás, Temperal, Temperaus, Os de cedo, Migueliña, Cabezuda	Spain	M	Late–very late	Complete	Low	Variable	M	Fresh, flour, marmalade, purée, timber
Tempestiva	S	Precoce di Roccamonfina	Italy	M	Early	–	–	–	–	Fresh
Temprana	S	–	Spain	L	Very early–early	Complete	High	Variable	L	Fresh
Temprano	S	–	Spain	XS	–	Free–Partial	Moderate	Thin	L	Different genotypes/Flour, marmalade, purée
Tempuriva	S	–	Italy	S–M	Early	Partial	Moderate	Thick	L	I(+)/Fresh
Tismana	S	–	Romania	M	Very late	–	–	–	–	High productivity, precocity
Tixera	S	–	Spain	XS	Medium–late	Free	Low	Very thin	M	Flour, marmalade, purée
Tomasa	S	–	Spain	XL	Medium–late	Complete	Very high	Thin	L	Fresh

(Continued)

Cultivar	Species	Synonyms	Origin	Nut Size	Ripening Time	Pellicle Adhesion	% of Doubles	Nut Stipes	Catkins	Remarks/Main Utilizations
Torbeana	S	Torbeana blanca Torbeana negra Torbeanas	Spain	XS	Late–very late	Partial	Nul	Variable	B	Flour, marmalade, purée
Torcione Nero	S	–	Switzerland	M	Medium	Free	Low	–	M	Fresh
Toubesa	S	–	Spain	L	Late–very late	Complete	Nul	Thin	B	Fresh, flour, marmalade, purée
Toyotamawase	C	–	Japan	S	Very early	Complete	–	–	B	–
Trigueira	S	–	Portugal	S	Late	Free	Moderate	Thin–thick	–	–
Tsuchidawase	C	–	Japan	L	Medium	Complete	Low	–	L	–
Tsukuba	C	–	Japan	XL	Early–medium	Complete	Low	–	L	Worldwide distributed
Tsunehisa	C	–	Japan	M	Very early–early	Complete	–	–	L	–
Uma só (Azores)	S	–	Portugal	L	Medium	–	–	–	–	–
Ünal	S	–	Turkey	S	Late	Free	Moderate	Thick	L	I(–)
Vakit Kestanesi	S	–	Turkey	XL	Early	Partial	–	–	B	I(+)
Valduna	S	–	Spain	M	Medium–late	Free	Low	Thin	L	Fresh
Vaquera	S	De Pie Vaquero	Spain	M	Medium–late	Partial	Very high	Thin–very thin	L	Fresh, flour, marmalade, purée

(Continued)

Cultivar	Species	Synonyms	Origin	Nut Size	Ripening Time	Pellicle Adhesion	% of Doubles	Nut Stipes	Catkins	Remarks/Main Utilizations
Vázquez	S	Vazqueña	Spain	L	Medium–late	Complete	Nul	Very thin	L	Fresh, candied
Vegamesada	S	Mesada	Spain	M	Medium–late	Partial	Moderate	Very thin	M	Fresh
Veiguiña	S	De Presa	Spain	S	Late–very late	Free	Nul	Thin–thick	M	Fresh, flour, marmalade, purée
Ventura	S	–	Spain	M	Late–very late	Free	Nul–moderate	Thin	M	Fresh, candied, flour, marmalade, purée
Verata	S	–	Spain	M	Late	Free	Moderate	Thin	A	Candied
Verdale	S	Delsol, Verdesa, Verdole	France	M	Late	Free–Partial	Nul	Thin	L	Pollinizer/Processed
Verde	S	–	Spain	S	Very late	Free	Low	–	A	Different genotypes/Flour, marmalade, purée
Verdeal	S	–	Portugal	L	Medium	Free	Moderate	Thick	–	–
Verdello	S	–	Spain	XS	Medium–late	Free	Very high	Thin	A	Flour, marmalade, purée
Verdesa	S	–	Switzerland	M	Late	Partial	Low	–	M	I(+)/Fresh
Verdesa (IT)	S	–	Italy	S	Late	Partial	–	–	L	Fresh
Vermelha	S	–	Portugal	M	Medium	Free	–	–	–	–
Viana (Azores)	S	–	Portugal	L	Late	Complete	Moderate	Thin	–	–

(Continued)

Cultivar	Species	Synonyms	Origin	Nut Size	Ripening Time	Pellicle Adhesion	% of Doubles	Nut Stipes	Catkins	Remarks/Main Utilizations
Vignols	S × C	–	France	L–XL	Medium–late	Free	High–very high	Thin	L	Worldwide distributed/I(–)/Fresh, candied, processed
Vileta	S	Courelá-Rubiá, Riá, Xabregá-Rubiá	Spain	M	Late	–	Moderate	Thin–very thin	L	Pollinizer/Flour, marmalade, purée
Villarenga	S	–	Spain	S	Late	Free	Low	Thick	M	Fresh, flour, marmalade, purée
Villaviciosa	S	–	Spain	S	Medium–late	Free	Moderate	Very thin	A	Flour, marmalade, purée
Vizcaína	S	Vizcaes	Spain	S	Medium–late	Free	Moderate	Thin–very thin	L	Flour, marmalade, purée
Volos	S	Voliotiko, Pelion	Greece	L–XL	Early–medium	Free	Low	Thin	–	I(–)/Fresh, candied, confectionary
Whitten North	S × D	–	USA	–	–	–	–	–	–	–
Whitten South	–	–	USA	–	–	–	–	–	–	–
Willamette	M × S	–	USA	L	–	Free	–	–	–	–
Wuhuali	M	–	China	S	Medium–late	Free	–	–	–	Cooked, roasted
Wukeli	M	–	China	M–L	Medium–late	–	–	–	–	Cooked

(Continued)

Cultivar	Species	Synonyms	Origin	Nut Size	Ripening Time	Pellicle Adhesion	% of Doubles	Nut Stipes	Catkins	Remarks/Main Utilizations
Xabrega	S	–	Spain	XS–S	Late–very late	Partial	Nul–low	Thin	L	Different genotypes/Flour, marmalade, purée
Xiaoyouli	M	–	China	S	Medium–late	–	–	–	–	Cooked, roasted
Xidra	S	Parede, Padana	Spain	XS	Medium–late	Free–Partial	Very high	Thin–very thin	M	Flour, marmalade, purée
Xilimendra	S	–	Spain	M	Late	Complete	Low	Thin	M	Fresh, flour, marmalade, purée
Xinyangdabanli	M	–	China	L	Very early–early	–	–	–	–	Cooked
Yakko	C	–	Japan	L	Very early–early	Free	Moderate	–	M	–
Yamaguchiwase	C	–	Japan	M	Very early–early	Complete	Moderate	–	M	–
Yamatowase	C	–	Japan	L	Very early–early	Complete	–	–	L	–
Yanchang	M	–	China	S	Early	–	–	–	–	Cooked, roasted
Yangmaoli	M	–	China	L	Late	Free	–	–	–	Cooked
Yanjin	M	–	China	M–L	Early	–	–	–	–	Cooked
Yankui	M	–	China	M–L	Very early–early	–	–	–	–	Cooked, roasted
Yankui	C	–	Japan	–	–	–	–	–	–	
Yanshanhong	M	–	China	S	Early	Free	Nul–low	–	–	Cooked, roasted

(*Continued*)

Cultivar	Species	Synonyms	Origin	Nut Size	Ripening Time	Pellicle Adhesion	% of Doubles	Nut Stipes	Catkins	Remarks/Main Utilizations
Yanshanzaofeng	M	–	China	S	Early	–	–	–	–	Cooked, roasted
Yeuilliaz	S	–	Italy	M	Medium	Partial	–	–	L	Fresh
Yinfeng	M	–	China	S	Medium–late	–	–	–	–	Cooked, roasted
Yixin	M	–	USA	–	–	–	–	–	–	–
Yoo Ma	C	–	USA	–	–	–	–	–	–	–
Yuluohong	M	–	China	M–L	Medium–late	–	–	–	–	Cooked

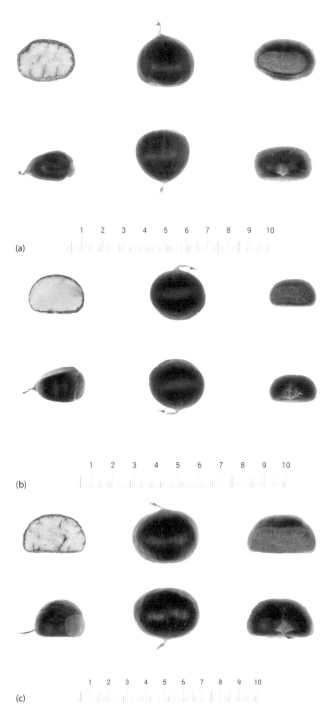

FIGURE 4.6   (a) Savoye; (b) Seguin's chestnut; (c) Solenga.

(*Continued*)

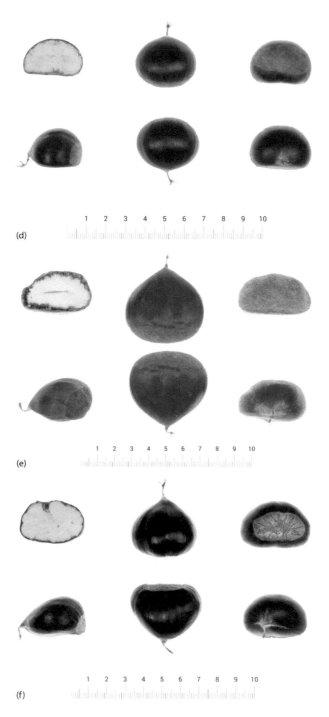

FIGURE 4.6 (Continued)   (d) Tarvisò; (e) Tsukuba; and (f) Volos. (a, c, d, e,: Courtesy of Betta, M.; f: Courtesy of Gamba, G.; b: Courtesy of Riondato, I.)

Cultivar	Species	Synonyms	Origin	Nut Size	Ripening Time	Pellicle Adhesion	% of Doubles	Nut Stipes	Catkins	Remarks/Main Utilizations
Yunyao	M	–	China	M–L	Very early	–	–	–	–	Cooked, roasted
Yunzao	M	–	China	M–L	Very early	–	–	–	–	Cooked, roasted
Za	M	–	China	S	Very early–early	–	–	–	–	Cooked, roasted
Zaolizi	M	–	China	L	Early	–	–	–	–	Cooked
Zeive	S	–	Portugal	L	Late	Free	–	–	–	–
Zhenandabanli	M	–	China	M–L	Very early–early	–	–	–	–	Cooked
Zhongchibanli	M	–	China	L	Medium–late	–	–	–	–	Cooked
Zhongchili	M	–	China	L	Medium–late	–	–	–	–	Cooked
Zhongguohongpili	M	–	China	M–L	Early	–	–	–	–	Cooked
Zocca	S	Marrone di Zocca	Italy	L	Medium	Free	–	–	A	I(–)/Fresh, marrons glacés, processed

Species: C = *Castanea crenata*; D = *Castanea dentata*; H = *Castanea henryi*; M = *Castanea mollissima*; P = *Castanea pumila*; S = *Castanea sativa*; Se = *Castanea seguinii*; Y = Complex hybrid; **Nut size:** XL = very big, <60 nuts/kg; L = big, 61–80 nuts/kg; M = medium, 81–100 nuts/kg; S = small, 101–120 nuts/kg; XS = very small, >121 nuts/kg; **Pellicle adhesion:** Free = pellicle not adherent; Partial = pellicle partially adherent; Complete = pellicle completely adherent; **% of double nuts** (percentage of nuts with more than 2 mm wrinkles): Nul = 0 multiple-embryo nuts; Low = 1–4 multiple-embryo nuts; Moderate = 5–8 multiple-embryo nuts; High = 8–12 multiple-embryo nuts; Very high = >12 multiple-embryo nuts; **Catkins:** A = astaminates; B = brachistaminates; L = longistaminates; M = mesostaminates; **Pellicle intrusion with (Remarks/main utilizations)** I(+) = presence of intrusions; I(–) = absence of intrusions.

4.2 BREEDING

Daniela Torello Marinoni, Santiago Pereira-Lorenzo,
Rita Lourenço Costa, and Roberto Botta

Breeding and selection are extremely important to obtain new and valuable cultivars for superior nut and timber production, to find and introduce resistance genes to main pest and diseases, to improve adaptability to different environments and to increase aptitude to vegetative propagation. Nowadays in Italy, Spain, France, Portugal, and Switzerland, old groves are being renovated and new orchards are growing. In North America, Australia, New Zealand, Chile, and Argentina new plantings are raising, increasing the necessity to obtain high yielding varieties and rootstocks adapted to a more intensive management of orchards. On the other hand, Asian production continues to increase, as both China and South Korea modernize their plantings and improve nut conservation techniques, in order to increase export markets especially towards Japan and the United States.

Breeding objectives are driven by final use (nuts or timber or rootstock) and market demands (fresh or processed uses). Concerning nut production, both fresh market and processing market require high quality chestnuts, large nut size, easy peeling, and suitability to storage and processing. For timber, important traits are wood quality, rapid growth, and non-checking wood (no ring-shake). In both cases, currently the primary objectives are ease of propagation and resistance to abiotic and biotic stresses, in particular to chestnut blight, ink disease and gall wasp; in modern orchards resistant rootstocks to the ink disease (*Phytophthora* spp.) are needed. The gene pool available for breeding includes the entire genus *Castanea*; in many cases, resistance sources to pest and pathogens are found in the Asiatic germplasm and interspecific hybrids.

4.2.1 PLANT CHARACTERISTICS

Modern chestnut cultivation is based on high density orchards and therefore semi-compact habitus, medium or low vigour are the most suitable plant traits. Other important traits are precocious bearing, regular and high yield, strong branches, good pollinizer ability, and inter-compatibility with the best cultivars. Moreover, an upright habit and low detaching force of the burs are important harvest-related traits for mechanical shaking. For timber production, cultivars with high vigour, high wood production, straight trunk, and wood not subject to ring-shake or radial checking are desired.[1] Moreover, some cultivars combine nut and timber quality suitable to be used in appropriate soils and environments.[2]

4.2.2 NUT CHARACTERISTICS

For the fresh market and for candying (marrons glacés), a large nut size is desirable, while for dried chestnuts and making flour a small or medium size nut is suitable. Evenness of shape, colour, dark brown stripes, easy pellicle removal, no pellicle intrusion, flavour are valuable traits for fresh marketing. The ease of peeling and small to medium size are also desirable nuts requirements for processing. Other essential traits, required both by fresh market and industry, are resistance or low susceptibility to pests (*Cydia, Curculio, Dryocosmus*) and fungi (*Gnomoniopsis castanea*).[1]

In France, among the genotypes selected for their resistance to ink disease, some ('Marigoule', 'Bournette', 'Maridonne', 'Bouche de Bétizac') exhibit desirable fruit quality traits.[3] In addition, following the introduction of gall wasp in Europe 'Bouche de Bétizac' was found to be resistant to this pest.[4] Moreover, in the frame of a breeding programme developed in France in the 1980s, ten promising nut cultivars were retained; among them, the cultivar for nut industry 'Bellefer®' shows very good processing qualities, great flavour, good crop potential, resistance to ink disease, and moderate sensitivity to canker and chestnut gall wasp (https://www.sival-innovation.com/en/bellefer-chestnut-variety/).[5] In Italy, 'Lusenta' and 'Primato' which show high yield, good quality nuts were obtained by hybridization between European and Japanese chestnuts.[6] Japanese chestnut breeding programs have focused on excellent cultivars with high nut quality and easy pellicle removal. The released cultivar 'Porotan' is the most prominent cultivar with good nut quality and an easy peeling trait for East Asian Markets[7] and, recently, the Japanese chestnut cultivar 'Yakko',[8] which had a different genetic origin, but also showing easy peeling and, additionally, a later nut harvest time than 'Porotan.'

4.2.3 ROOTSTOCKS AND PROPAGATION

Good aptitude for vegetative propagation and rootstock/scion compatibility are of primary importance in rootstock breeding. *C. crenata* hybrids (*C. crenata* × *C. sativa* and *C. sativa* × *C. crenata*) are easier to propagate by cuttings or layering than *C. Sativa*.[9,10] The most popular clonal rootstocks are the Eurojapanese hybrids selected in France and include: 'CA 07' ('Marsol') (moderately resistant to *Phytophthora*); 'CA 74' ('Maraval') (*Phytophthora* resistant, low vigour); 'CA 118' ('Marlhac') (moderately resistant to *Phytophthora*, but able to grow at temperature $< -10°C$); 'CA 90' ('Ferosacre') (*Phytophthora* resistant, but sensitive to temperatures $< -10°C$). Unfortunately, graft incompatibility problems with many European cultivars have limited their wider use.[9,11–13]

There is an urgent need of rootstocks adapted to specific ecological conditions and compatible with *C. sativa* cultivars. For example, in mountainous European southern regions, edaphic and climatic conditions strongly limit the use of hybrid clones; considering the large amount of variation in ink disease resistance in the European species.[14,15] A new program is being carried out by INRA to select new and suitable rootstocks within *C. sativa*.

In Portugal, the new released rootstock resistant to ink disease ColUTAD® is being broadly evaluated in different growing areas to test the suitability with different European cultivars.[7]

In Portugal new rootstocks, with improved resistance to *P. cinnamomi*, were selected from the breeding program, initiated in 2006, based on controlled crosses between the European chestnut (*C. sativa*) and the Asian species, *C. crenata* (the Japanese chestnut) and *C. mollissima* (the Chinese chestnut) used as donors of resistance. The new rootstocks are being multiplied by micropropagation and cuttings, and are being grafted with European chestnut varieties. A field test was installed with these new rootstocks, where the performance of each one is being evaluated in terms of adaptation, vigour and compatibility of grafting with four Portuguese varieties ('Longal', 'Martaínha', 'Judia' and 'Bária'). They are also being characterized for 39 descriptors UPOV and they will be released to the market in 2020/2021. The aim is to supply the market with new improved rootstocks, better adapted to the actual conditions of climate, since the ones commercialize nowadays in Portugal come from the breeding programs of the 1950s and 1960s and show poor adaptation in certain production regions.

4.2.4 RESISTANCE TO ABIOTIC AND BIOTIC STRESSES

Abiotic: resistance to spring frost is especially important for Eurojapanese hybrids, which have an early leaf budburst and flowering time. Resistance to drought conditions is desirable, especially for interspecific hybrids, to expand chestnut cultivation into temperate, warm, and dry zones.[1]

Biotic-diseases: Two main diseases, ink disease (*Phytophthora* spp.) and blight (*Chryphonectria parasitica*), threaten chestnut production. The European chestnut species *C. sativa* present less tolerance to main diseases than Asian species.

Ink disease (Phytophthora *spp.*): Ink disease causes serious damages in Europe, and also in China, Japan, Turkey, and the United States. Resistances to ink disease have been found in *C. mollissima* and *C. crenata*.[14,16-18] However, the introduced genotypes were not adapted to European or American environmental conditions and did not match growers' and consumers' demands. Therefore, chestnut breeding in Europe began with the production of hybrids resistant to ink disease (*Phytophthora* spp.). In 1989 a new program began to identify some hybrid clones that were interesting for timber, nut production, or rootstocks.[7,13] In France, the first deliverables of the chestnut breeding program were *C. crenata* × *C. sativa* cultivars: 'Marsol', 'Maraval', 'Marigoule' and 'Précoce Migoule.' Selected first as ink disease resistant varieties for timber production, they were used, very quickly after their release, as rootstocks and become very popular as such, in several countries.

The first interspecific hybridizations in Portugal were initiated in 1947 by Bernardino Barros Gomes, in order to introduce resistance to ink disease in *C. sativa*.[19,20] Resistant *C. sativa* selections from ColUTAD® in Portugal are being tested in a micropropagation program in order to rapidly make them available to the producers.[7]

New rootstocks with improved resistance to *P. cinnamomi*, selected from the breeding program initiated in 2006, will be released in 2020/2021.[21,22]

Chestnut blight (Chryphonectria parasitica *(Murr.) Barr*): One of the main objectives of chestnut breeding is to improve resistance to the fungal pathogen *C. parasitica*. The primary source of resistance to chestnut blight is coming from Asian species, mainly *C. mollissima*.[23] In 1981, the American Chestnut Foundation initiated a new breeding program to introgress disease resistance into *C. dentata* populations, with the goal to restore American chestnut with hybrids that retain desirable American chestnut phenotype, but resistant to blight.[24] Twenty five years later, selected genotypes from three back-cross generations proved to be resistant with similar morphological characteristics to that of *C. dentata*[25] and the first forest reintroduction trials were established in 2009.[26,27]

Biotic–pest: insect species that damage chestnut and can be serious limiting factors in production are chestnut gall wasp (*Dryocosmus kuriphilus* Yasumatsu), nut weevil (*Curculio elephas* L.), lepidote moths (*Pammene* spp. Hb. and *Cydia* spp. Hb.), and the wood-boring beetle (*Xyleborus dispar* F.).[1]

Chestnut gall wasp (Dryocosmus kuriphilus *Yasumatsu*): The use of cultivars bearing resistance, or low susceptibility, to the pest, combined with the use of biological control by natural parasitoid *Torymus sinensis* Kamijo has been successful in different parts of the world and have contributed to overcome the emergency of the infestation and yield losses. Yet, studies on host-pathogen interaction and the search for strategies based on quantitative and qualitative resistance are still of relevant interest for breeding programs.

In Japan, *C. crenata* resistant cultivars 'Tanzawa', 'Ibuki,' and 'Tsukuba' were bred in 1959, followed by 'Ishizuki' in 1968 and allowed a recovery of the chestnut industry.[28] However, these resistances broke down, and two cultivars with better resistance to gall wasp were released in 1981 ('Kunimi') and 1992 ('Shihou'). Other breeding strategies, adopted in Korea and China, were based on the selection of cultivars resistant to the cynipid by focusing on some varieties of *C. mollissima* and hybrids *C. crenata* × *C. mollissima*, with late bud formation. In this case, at the time of adult emergence, the cynipid does not find buds of appropriate size for the oviposition.[29] In the USA, renewed efforts have been made to breed chinquapins (*C. pumila* var. *pumila* and *C. pumila* var. *ozarkensis*) resistant to gall wasp bearing suitable characteristics for chestnut industry and for forest plantings.[30] Following the introduction of the cynipid in Italy in 2002, a study aimed at evaluating the susceptibility of local and international *C. sativa* and hybrid cultivars under controlled conditions was carried out starting in 2004 until 2013. Observations showed large differences in terms of level of infestation among cultivars, and the occurrence of resistance in *C. sativa* cultivars and Eurojapanese hybrids, including 'Bouche de Bétizac', 'Marlhac' and 'Maridonne' (all *C. sativa* × *C. crenata*).[31]

Carpophagous insects: Other two main insect pests are the moth larvae *Cydia* (=*Laspeyresia*) *splendana* Hb. and the weevil *Curculio* (=*Balaninus*) *elephas* Gyll. It has been commonly observed that there is variation among trees in the degree to which they are infested, but no selection has been made. Debouzie *et al.* (1996)[32] demonstrated that presence of chestnut moth larvae inhibited weevil egg-laying. It appears that *Curculio* oviposits less in those cultivars with longest bur spines,[33] suggesting the length of the spines on the burs may be involved in resistance.

4.2.5 PRESENT AND FUTURE OUTLOOK

Selecting chestnut cultivars and rootstocks resistant to pest and diseases, easy to be clonally propagated, suitable to different environmental conditions, is an urgent issue which is currently addressed. Research in chestnut breeding resulted in a high number of registered rootstocks and nut varieties. The release of rootstocks resistant to ink disease has allowed to maintain or to increase the yield in several regions. Yet, chestnut blight and gall wasp continue to cause losses in chestnut production, but resistant nut varieties could be used in order to achieve a more complete control. Progress made by American researchers in understanding the mechanisms of resistance to chestnut blight is outstanding.[34,35] A better knowledge of defence mechanisms targeted to gall wasp[36] could contribute to breed resistant varieties; similarly, improving resistance of nuts to rots caused by several fungi (*Gnomoniopsis*, *Phomopsis*, *Sclerotinia* spp.) and to chestnut weevil or to carpophagous insects can improve chestnut cultivation.

In the future the ongoing climate change is one of the main challenges that chestnut will have to face and breeding is needed to reduce the effects of the changes; an increase of mean temperatures is expected in temperate regions and this will impact both on the phenology and on the water use efficiency of the trees, highlighting the need for stress tolerant cultivars. Some studies have demonstrated the variability of adaptive traits in chestnut showing a broad resilience to climatic changes, which can be exploited to adapt the species to new areas.[37]

The traditional methods used in chestnut for breeding are selective breeding and crossbreeding, but recent developments in the field of tree genome mapping and sequencing open up new perspectives for breeders with different applications, which rely on knowledge of the genome.[35] A way to improve chestnut breeding is the development of marker-assisted selection (MAS) for the creation of either nut cultivars or rootstocks, producing high quality fruits and resistant to abiotic and biotic stresses. Moreover, development of linkage maps[38] and genome-wide association studies (GWAS) will identify SNPs and genes related to phenotypic differentiation as in other fruit trees such as apple.[39] Genome-wide sequence information and functional genomics will establish the basis for crop genome engineering to modify and or include agronomical traits with unprecedented precision, keeping most of the original cultivars, by using genome editing and in particular the CRISPR/Cas9 technology.[40]

REFERENCES

1. Bounous, G. and Marinoni, D.T. 2005. Chestnut: Botany, horticulture, and utilization. *Horticultural Reviews* 31: 291–347.
2. Pereira-Lorenzo, S. and Ramos-Cabrer, A. 2004. Chestnut, an ancient crop with future. In *Production Practices and Quality Assessment of Food Crops*, Volume 1, 105–161. Springer, Dordrecht, the Netherlands.
3. Breisch, H. 1995. Châtaignes et marrons. CTIFL, Paris, France, p. 239.
4. Sartor, C., Botta, R., Mellano, M.G., Beccaro, G.L., Bounous, G., Torello Marinoni, D., Quacchia, A., and Alma, A. 2009. Evaluation of susceptibility to *Dryocosmus kuriphilus* Yasumatsu (Hymenoptera: Cynipidae) in *Castanea sativa* Miller and in hybrid cultivars. *Acta Horticulturae* 815: 289–297.

5. Pasquet, N. 2016. Point sur les Nouvelles Sélections Variétés et Porte-greffes. *Rencontres Techniques Châtaigne. Congress presentaion, Ctifl, Lanxade.*

6. Bounous, G. 2014. *Il Castagno: Risorsa Multifunzionale in Italia e nel mondo.* Edagricole, Bologna, Italy, p. 420.

7. Pereira-Lorenzo, S., Costa, R.M., Ramos-Cabrer, A.M., Ciordia-Ara, M., Marques Ribeiro, C.A., Borges, O. and Barreneche, T. 2011. Chestnut cultivar diversification process in the Iberian Peninsula, Canary Islands and Azores. *Genome* 54: 301–315.

8. Takada, N., Yamada, M., Nishio, S., Kato, H., Sawamura, Y., Sato, A., Onoue, N. and Saito, T. 2018. The investigation of pellicle peelability on Japanese chestnut cultivar of 'Yakko' (*Castanea crenata* Sieb. et Zucc.). *Scientia Horticulturae* 234: 146–151.

9. Chapa, J., Chazerans, P. and Coulie, J. 1990. Multiplication végétative du châtaignier, amelioration par greffage de printemps et bouturage semi-légneux. *L'Arboriculture Fruitierè* 431: 41–48.

10. Bounous, G., Bouchet, M. and Gourdon, L. 1992. Reconstruction of traditional chestnut orchard. *Experiences in Piedmont and Southern France. L'Informatore Agrario* 9: 155–160.

11. Ferrini, F., Mattii, G.B., Nicese, F.P. and Pisani, P.L. 1992. Investigation for realisation of clonal rootstock for chestnut trees. *Proceedings of a Scientific Meeting SOI*: 412–413.

12. Breisch, H. 1992. Compatibility tests between the main French varieties of chestnut trees and ink-resistant hybrid rootstocks. *Proceedings of the World Chestnut Industry Conference, Morgantown, WV. Chestnut Marketing Associate, Alachua, FL*, 41–53.

13. Pereira-Lorenzo, S. and Fernandez, J. 1997. Propagation of chestnut cultivars by grafting: Methods, rootstocks and plant quality. *Journal of Horticultural Science and Biotechnology* 72(2): 731–739.

14. Schad, C., Solignat, G., Grente, J. and Venot, P. 1952. Recherches sur le châtaignier à la Station de Brive. *Annales de l'Amélioration des plantes* 3: 369–458.

15. Robin, C., Morel, O., Vettraino, A.M., Perlerou, C., Diamandis, S. and Vannini, A. 2006. Genetic variation in susceptibility to *Phytophthora cambivora* in European chestnut (*Castanea sativa*). *Forest Ecology and Management* 226: 199–207.

16. Anagnostakis, S.L. 2012. Chestnut breeding in the United States for disease and insect resistance. *Plant Disease* 96(10): 1392–1403.

17. Clapper, R.B. 1954. Chestnut breeding techniques and results. *Journal of Heredity* 45(4): 201–208.

18. Dufrenoy, J. 1930. La lutte contre la maladie des châtaigniers. *Ann. Epiph.* 1: 3–49.

19. Gomes Guerreiro, M. 1948. Acerca do uso da análise discriminatória: Comparação entre duas castas de castanhas. *Sep. Das Publicações da Direcção Geral dos Serviços Florestais e Aquícolas* XV(I-II): 137–151.

20. Gomes Guerreiro, M. 1957. *Castanheiros: Alguns estudos sobre a sua ecologia e o seu melhoramento genético.* Instituto Superior de Agronomía, Lisboa, Portugal.

21. Costa, R., Santos, C., Tavares, F., Machado, H., Gomes-Laranjo, J., Kubisiak, T. and Nelson, C.D. 2011. Mapping and transcriptomic approaches implemented forunderstanding disease resistance to *Phytophthora cinnamomi* in *Castanea* sp. *IUFRO Tree Biotechnology Conference 2011: From Genomes to Integration and Delivery* 5(Suppl 7): A1.

22. Santos, C., Machado, H., Correia, I., Gomes, F., Gomes-Laranjo, J. and Costa, R. 2015. Phenotyping *Castanea* hybrids for *Phytophthora cinnamomi* resistance. *Plant Pathology* 64: 901–910.

23. Hebard, F.V. and Stiles, S. 1996. Backcross breeding simplified. *Journal of the American Chestnut Foundation* 10: 35–39.

24. Burnham, C. 1981. Blight resistant American Chestnut: There's hope. *Plant Disease* 65(6): 459–460.

25. Diskin, M., Steiner, K.C. and Hebard, F.V. 2006. Recovery of American chestnut characteristics following hybridization and backcross breeding to restore blight-ravaged *Castanea dentata*. *Forest Ecology and Management* 223: 439–447.

26. Clark, S.L., Schlarbaum, S.E., Saxton, A.M. and Hebard, F.V. 2014. The first research plantings of third-generation, third-backcross American Chestnut (*Castanea dentata*) in the Southeastern United States. *Acta Horticulturae* 1019: 39–44.

27. Clark, S.L., Schlarbaum, S.E., Saxton, A.M. and Hebard, F.V. 2016. Establishment of American chestnuts (*Castanea dentata*) bred for blight (*Cryphonectria parasitica*) resistance: Influence of breeding and nursery grading. *New Forests* 47(2): 243–270.

28. Pereira-Lorenzo, S., Ballester, A., Corredoira, E., Vieitez, A.M., Agnanostakis, S., Costa, R., Bounous, G., Botta, R., Beccaro, G.L. and Kubisiak, T.L. 2012. Chestnut. In *Fruit Breeding*, pp. 729–769. Springer, New York.

29. Norton, J.D., Harris, H. and Conaty, T.E. 1987. Resistance to chestnut gall wasp in Chinese chestnut. In: Chestnuts and creating a commercial chestnut industry. *Proceedings of the Second Pacific Northwest Chestnut Congress*, 30–32. Chestnut Growers exchange, Portland, OR.

30. Anagnostakis, S.L. 2014. A preliminary report on Asian chestnut gall wasp on species and hybrids of chestnut in Connecticut. *Acta Horticulturae* 1019: 21–22.

31. Sartor, C., Dini, F., Torello Marinoni, D., Mellano, M.G., Beccaro, G.L., Alma, A., Quacchia, A. and Botta, R. 2015. Impact of the Asian wasp *Dryocosmus kuriphilus* (Yasumatsu) on cultivated chestnut: Yield loss and cultivar susceptibility. *Scientia Horticulturae* 197: 454–460.

32. Debouzie, D., Heizmann, A., Desouhant, E. and Manu, F. 1996. Interference of several temporal and spatial scales between two chestnut insects. *Oecologia* 108: 151–158.

33. Bergougnoux, F., Verlhac, A., Breisch, H. and Chapa, J. 1978. Le Châtaignier Production et Culture. *Journées nationales du chataignier*. Nîmes, France: Comité national interprofessionnel de la chataigne et du marron.

34. Worthen, L.M., Woeste, K.E. and Michler, C.H. 2010. Breeding American chestnuts for blight resistance. *Plant Breeding Reviews* 33: 305–335.

35. Nelson, C., Powell, W., Merkle, S., Carlson, J.E., Hebard, F.V., Faridi, N., Staton, M.E. and Georgi, L. 2014. Chestnut. In *Tree Biotechnology*, 3–35. CRC Press, Boca Raton, FL.

36. Dini, F., Sartor, C. and Botta, R. 2012. Detection of a hypersensitive reaction in the chestnut hybrid 'Bouche de Bétizac' infested by *Dryocosmus kuriphilus* Yasumatsu. *Plant Physiology and Biochemistry* 60: 67–73.

37. Ciordia, M., Feito, I., Pereira-Lorenzo, S., Fernández, A. and Majada, J. 2012. Adaptive diversity in *Castanea sativa* Mill. half-sib progenies in response to drought stress. *Environmental and Experimental Botany* 78: 56–63.

38. Torello-Marinoni, D., Nishio, S., Portis, E., Valentini, N., Sartor, C., Dini, F., Ruffa, P. et al. 2018. Development of a genetic linkage map for molecular breeding of chestnut. *Acta Horticulturae* 1220: 23–28.

39. Leforestier, D., Ravon, E., Muranty, H., Cornille, A., Lemaire, C., Giraud, T., Durel, C.E. and Branca, A. 2015. Genomic basis of the differences between cider and dessert apple varieties. *Evolutionar Applications* 8: 650–661.

40. Osakabe, Y. and Osakabe, K. 2015. Genome editing with engineered nucleases in plants. *Plant and Cell Physiology* 5: 389–400.

5 Nursery Techniques

*Gabriele Beccaro, Giancarlo Bounous,
Beatriz Cuenca, Michele Bounous, Michele
Warmund, Huan Xiong, Li Zhang, Feng Zou,
Ümit Serdar, Burak Akyüz, Maria Gabriella
Mellano, and Rita Lourenço Costa*

CONTENTS

5.1 INTRODUCTION

The propagation method is often determined by the intended uses of the tree. For forestal use, sexually propagated trees should be used, and the seeds will have to fulfill the legal requirements for this use (i.e. seeds sh ould come from the same provenance where the forestation is done). In Europe, for example, the Directive 1999/105CE requires the use of forestal materials including chestnut, and this directive has been applied to each country. For wood production in some countries, clonal Euro-Japanese hybrids are often used because of their vigorous growth. For nut production, grafting a nut cultivar scion onto a rootstock is necessary. For this purpose, seedling rootstocks are widely used; however, clonal rootstocks have some advantages: tree uniformity, regular shape of the trees, better adaptability of the tree to soil conditions, with tolerance to diseases (ink disease for example), or enhanced graft compatibility with a specific cultivar. These clonal rootstocks, however, are produced by rooting shoots from a superior clone that possess a specific characteristic (tolerance to diseases, to poor soils, grafting compatibility, etc.) and as only plants in a juvenile phase are able to root; the donor clone must be kept in a juvenile stage so its rooting ability is not lost.

5.2 SOWING SEEDS FOR ROOTSTOCKS AND SEED TREES

5.2.1 SEEDS CHOICE AND CARE

Chestnut seedlings are used for forestation, wood production, and as rootstocks for grafting cultivars. Propagation via seed also often occurs with *C. dentata*, chinqua-pins, minor *Castanea* species, and ornamental cultivars. The best seeds are well-formed, of medium size, with low percentage of pellicle intrusion or with a single embryo to obtain only one seedling per nut. Seeds must be healthy and, to avoid mold development, picked up immediately after they drop to the ground.

It is preferable that the seeds come from plants located in environments similar to those of the cultivar to maximize the grafting compatibility between rootstock and cultivar. It is better if rootstocks and scions belong to the same species. Euro-Japanese hybrids can be grafted on *C. sativa* or *C. crenata* seedlings (and their intraspecific open pollinated seedlings) or hybrid clonal rootstocks.

Nut conditioning for germination should be done as soon as possible after the collection. Plant debris and burs should be removed before immersion in water. Immature, undersized, and parasitized nuts float in water and are discarded. The following practices are used to prevent insect infestation and prolong storage of chestnuts[1]:

- Immersion in hot water (48°C–50°C) for 30–45 minutes, followed by an immersion in cold water for 8–12 hours, to recover the initial temperature and to avoid heating processes;
- Immersion in cold water for 5–8 days (using a 2 chestnut: 3 water ratio relation chestnut/water 2:3), to create an anaerobic condition with a double effect: to eliminate the aerobic microorganisms causing rot and mold, and to encourage the development of anaerobic microorganism promoting a lactic fermentation that decreases the sugar concentration, therefore reducing the nutritive substrate for rotting fungi.

Seeds cannot be stored for a long time, not even refrigerated, as they will lose the germination ability. Chestnut seed can be stored for a short period (few months) by sealing them in a polyethylene bag, at 4°C. Alternatively, they can be stored at

50%–55% R.H. (relative humidity) at 0°C–3°C for overwinter storage, or slightly lower (−3°C–0°C) if a longer storage is needed. The storage can only be done in open containers in storage chambers with the established temperature and relative humidity between 90%–95%. When the environment is not humid enough, containers that ensure humidity as well as gas exchange can be used.[2]

Good germination occurs after an adequate cooling in humid conditions to remove growth inhibitory hormones. As already stated, to avoid dehydration which reduces germination, chestnuts are stratified in sand or in humid peatmoss at 1°C–2°C for at least 2 months until 4 or 5 months, depending on the time of sowing. After stratification, germination occurs in a more synchronized way, so a more uniform lot of plants is obtained. During germination, chestnut seeds absorb high quantities of water and O_2 and stored nutrients in the cotyledons are transferred to the growing embryo. The primordial root (radicle and hypocotyl) and the young shoot (epicotyl) both emerge from the apex of the nut (torch).

Pre-germinated nuts are sown in raised beds, trays, or in pots. The soil substrate should be soft and well drained (Figures 5.1 and 5.2).

The germinated chestnuts are sown with the flat side down to facilitate root penetration in the soil. The depth of seeding must not exceed 3–5 cm and, if sown in a raised bed, spacing in the row is usually 30–40 cm with rows 80–100 cm apart. At the end of the first growing season (August–September), raised beds seedlings are 100–150 cm tall with a diameter of 8–12 mm and are ready to be chip budded in fall or grafted or budded the following spring. However, chestnut container or tray production is becoming more common, allowing flexibility for sowing dates, and better conditions to control germination and plant development. In this case, sowing is normally done inside the greenhouse so it can be done earlier in winter. The germination process occurs under protected conditions, and in spring,

FIGURE 5.1   *C. dentata* seedling. (Courtesy of Beccaro, G.)

FIGURE 5.2   *C. mollissima* seedling. (Courtesy of Beccaro, G.)

FIGURE 5.3   Chestnut seedling raised in 400 cm³ forestal trays. (Courtesy of Cuenca, B.)

chestnut seedlings are moved outdoors after there is no risk of frost. Pots should be big enough (more than 400 cm³), to allow good root development, and the resulting plant attains an adequate size and quality (Figure 5.3).

Substrates normally used are a mix of peat with vermiculite or coco fiber. Pine bark is no longer recommended due to the incidence of *Fusarium circinatum* fungi. In these conditions, germination percentage is normally 80%. After germination, plant density is important. If it is too high, shading occurs and overlapping plant leaves restrict water movement through the growing medium, resulting in poor growth, dehydration, or plant mortality. A density of 100 plants per m^2, using a pot size of 600–1000 cm^3, allows adequate space for plant growth.[2]

In the USA, seeding rootstocks are produced by air root pruning to induce a more highly branched, fibrous root system. After nuts are stratified in sealed polyethylene bags, they are placed in a shallow flat with a 1.6 cm square bottom mesh for air root pruning on a wire bench in a greenhouse. After the primary root grows to the bottom of the container and is exposed to air, the root tip desiccates and lateral roots develop. Just after primary root desiccation, each seedling is transplanted into an individual container with an open, cross banded-bottom for further growth and more air root pruning in a greenhouse before grafting.

5.2.2 Care for Seedlings

Irrigation, fertilization, and phytosanitary care are very important for seedlings. In the field, hoeing or mechanical weed control is used to avoid competition for soil moisture and nutrients. Plastic mulch, can also be used to inhibit weed seed germination, but is not an environmentally sustainable practice and gives lot of work to the nursery. In large nurseries, herbicides are used for weed control (see Chapter 6). Seedling rootstocks are grown with a single stem. Grafting is performed in the autumn or in the following spring, when the seedling diameter is of sufficient size for grafting.

5.2.3 Direct Sowing in Forest

Direct sowing in forest or orchard was once used to increase the number of chestnut trees or to replace failures. Today, planting grafted trees in these situations is preferable to sowing seeds for several reasons. Seedling establishment in the field is difficult, because of herbivores and the high risk of canker blight infection after grafting. Because of these problems, a heavy seeding rate should be used when seeding in the forest. In any case, if canker blight is not present and it is necessary to apply this technique in forest or in open field, it is important to sow at least 2–3 seeds due to the high failure risk. After the emergerence, they must be thinned, adequately supported, and protected with a shelter against hares or other wild animals such as deer. *In situ*, fertilization, frequent irrigation and mulching are necessary. Seedling rootstocks grown in the field are generally suitable for grafting after one or two years. In the forest, it is generally preferable to graft on pollarded coppices (see Chapter 8).

5.3 PRODUCING CLONAL ROOTSTOCKS OR SELF-ROOTED TREES VIA VEGETATIVE PROPAGATION

Layering, rooting soft, or hardwood cuttings, and micropropagation are techniques used to produce self-rooted rootstocks, or self-rooted chestnut trees primarily for *C. mollissima*, although it is not a widespread practice. As the demand for clonal plants is rapidly increasing, there is a need for innovation in the chestnut nurseries. Research and private nurseries focused on the development of innovative chestnut propagation techniques should strengthen traditional nursery systems.

Clonal rootstocks that impart dwarfing or tolerance to *Phytophthora*, are needed to revitalize chestnut cultivation, using planting systems with a uniform layout and cultural practices that facilitate modern field operations.

Chestnut orchards are incorporating the same technologies as modern fruit orchards, using trees with a disease-resistant clonal rootstock and precocious scion cultivar that bear nuts at a young tree age. Thus, vegetative production methods are essential for chestnut rootstock production, and for the modernization of the chestnut production sector.

5.3.1 LAYERING

Most Eurojapanese genotypes can be propagated by layering (Technical Sheet 5.1). This method presents some advantages: it is easy to perform, inexpensive (no need for high technology or complicated facilities), it does not require highly skilled labor, and it can produce plants within a year. However, it has also some disadvantages: sufficient land is needed when producing many; there is a high demand of labor concentrated in a short time; the root system produced may be poorly developed; plants can only be produced during the dormant period; and any change in the clones under production is difficult and expensive.[2]

TECHNICAL SHEET 5.1　LAYERING

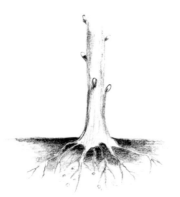

1. Mother plant   2. The plant is cut 10 cm from the ground.

3. Sprouts grow from the stump.

4. Soil is mounded and stems start rooting.

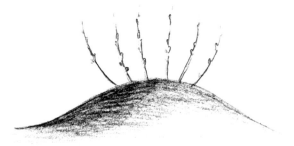

5. In autumn the shoots lose their leaves and are fully rooted.

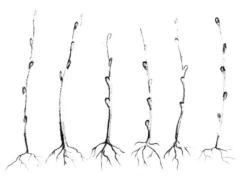

6. Rooted shoots are separated from the mother plant.

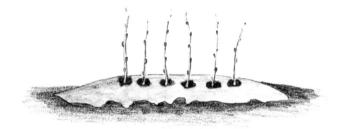

7. The shoots are transplanted.

Courtesy of Facello, V.

Mound layering or stooling is used to develop new plants by rooting stems attached to the mother plant.[3] Mother plants are spaced 1–1.2 m within the row and 2.5 m between rows. In the third year, the mother plant is cut at about 10 cm from the soil surface to promote new growth from the stems. The most juvenile parts of a tree, and therefore the ones able to root, are those located closest to the root. This is why the mother plant is cut almost to the ground level (Figure 5.4a). When sprouts grow from the stump, from the end of May to June (Figure 5.4b), the best stems are selected (those of medium diameter, healthy and upright) and defoliated in the basal part, while lateral, large, or small stems are eliminated. The selected stems are girdled by a thin wire 2–5 cm above their base (Figure 5.4c).

Girdling induces hyperplasia and enzymatic activity at the base of the stems. This practice promotes the accumulation of IAA oxidase, peroxidase, and polyphenol oxidase after 40–50 days. After girdling, the portion of the stem below the wire is scratched and painted with a mastic made with Vaseline® (petroleum jelly) or some other

(a)   (b)   (c)

FIGURE 5.4   Stooling process: (a) mother plant beginning to sprout in spring; (b) mother tree fully sprouted; (c) mother tree with cleared out, girded and hormoned shoots. (Courtesy of Míguez M.)

(a)

(b)

(c)

FIGURE 5.5 Stooling process: (a) covering of the tied-up mother plants; (b) uncovering and harvesting of the new rooted plants; (c) ready to sell plants in the nursery. (Courtesy of Míguez M.)

carrier for IBA, usually at a concentration of 4%–8%. Then the stems are tied together to avoid breaking when covered. The stem and the stump are covered with loose, acidic (pH 4.5–5) soil and kept moistened (Figure 5.5a). Etiolation modifies the anatomic structure and the physiological mechanism, favoring rooting. Rooting usually occurs after 60–70 days. If the pH is higher, the percentage of rooting decreases.[3] Weeding, irrigation, and NPK fertilization are needed. Digging and harvesting the newly rooted plants are done from after defoliation in mid-November (Figure 5.5b and c).

This propagation method commonly results in unbalanced rooting systems and a curve in the base of the stems where the plant was joined to the mother plant. Both faults can be corrected with one more year in the nursery, cutting the malformed stem, and choosing a new upright shoot next spring. This second year growth also results in a larger, better-developed root system.

5.3.2 Cutting

It was traditionally thought that chestnut was a difficult species to root by cuttings. However, the development of humidification systems has changed this, allowing rooting of chestnut stems that are not lignified (semi-herbaceous). In the last decades, the Forest Research Centre of Lourizán (CIF Lourizán) in Spain has developed the semi-herbaceous cutting methodology, which allows the rooting of cuttings using substrates that ensure adequate drainage after their basal immersion in a concentrated auxin solution.[4]

Rootstock propagation from cuttings is currently used by some commercial nurseries in Spain, Portugal, France, and Italy. Some of the advantages of this system are a better quality root system; more cuttings can be obtained per mother plant compared with stooling; a more controlled process; and as plants after rooting are normally potted, the nursery has more time to sell them. Nevertheless, this technique is not commonly used for mass propagation for industry purposes, due to some disadvantages such as the need for specialized facilities with humidity control; plants may exhibit plagiotropism; poor branching of root system compared to seedlings; change in clones under production is problematic and expensive; and the intensive labor requirement is concentrated in a very short time.[2]

5.3.2.1 Mother Plants

Mother plants are young or juvenile plants subjected to severe pruning to obtain scion wood used to prepare cuttings in the spring. These plants can be grown in the field or in pots in greenhouse (Figure 5.6a), ensuring better sanitary control. Rooting efficiency strongly decreases in cuttings collected from mother plants older than 5 years. Biochemical, anatomical, and environmental factors influence adventitious

(a)   (b)

(c)   (d)

FIGURE 5.6 Cutting process: (a) a heavily pruned, potted mother plant beginning to sprout; (b) fully sprouted mother plants in June; (c) collection of the shoots; (d) cuttings prepared for propagation. (Courtesy of Cuenca, B.)

rooting; in general, young plants have higher quantities of root promoters and lack auxin-inhibiting factors. Juvenility, etiolation, exogenous promoters, and mycorrhizal fungi promote rooting. Thus, the success of cutting from young plants is higher than from adult plants.[5] However, mother plants can be kept in a juvenile stage to maintain their ability to root. Juvenility gradient of a meristem is inversely proportional to the distance (along the branches and trunk) between the root collar and the meristem.[6] Therefore, cuttings collected from severely pruned plants will be ontogenically young and maintain their ability to root, although coming from physiologically old mother plants. Mother plants are certified true-to-type and free from chestnut mosaic virus (ChMV), and substrate composition strongly influences their growth and efficiency.[7]

The success of propagation by softwood cutting depends not only on genotype and stem juvenility, but also on the time of cutting collection and the use of growth regulars. Other factors such as soil, relative humidity, and temperature are also important. *C. crenata* × *C. sativa* hybrids and *C. mollissima* have a high capacity for self-rooting. However, also cuttings of some *C. sativa* genotypes, when taken from young mother plants (i.e. by micropropagation), may root easily in culture.[8] In Spain, the company TRAGSA roots around 50,000 cuttings/year of Euroasiatic hybrids for use as clonal rootstocks.[9,10] In Italy, the Chestnut Research & Development Center Piemonte has also developed an effective method for rooting Eurojapanese clonal rootstocks, with 60% to 80% success.[11]

5.3.2.2   Cutting Treatments

Semi-herbaceous stems are collected around June from mother plants grown in large pots (Figure 5.6b and c). Next, they are cut into 1 to 3 bud sections, about 7–15 cm-long, with one bud located at the base of the cutting and one at the top. The apical leaf or at least part of it is left attached to the cutting (Figure 5.6d).

Because buds and leaves are sites of auxin synthesis, so adventitious roots develop from the buried basal bud. The basal end of cuttings are dipped in a IBA solution (1000–4000 ppm) before placing them in rooting containers trays filled with peat/perlite/bark media inside climate-controlled rooms or rooting tunnels in the greenhouse (Figure 5.7a and b). A fog system, inside the tunnel, ensures high humidity. In the first week, the R.H. is maintained around 100% to avoid desiccation, but it is reduced incrementally. The growth medium temperature is around 23°C–24°C. Basal heating with propagation mats or hot water pipes is helpful to maintain a minimum temperature of 15°C at night and to shorten the rooting period. During the rooting process, plants continue their development, the internode length increases, young leaves grow, and first roots appear after 50–60 days (Figure 5.7c).[3] In the C-box system, cuttings are grown in climatic cells and in plastic bags[11] (Figure 5.8).

For some cultivars, rooting is challenging. The presence of growth inhibitors in adult plants interferes with indoleacetic acid (IAA). Hydroxybenzoic, vanyllic, salicylic, siryngic acids, and some water-soluble compounds also reduce rooting. Treatments with fungicides are necessary to avoid the risk of rot during the rooting process. Young rooted plants also require foliar fertilization, mainly with nitrogen. The following spring, young rooted plants are transplanted in the field or in pots to increase their growth (Figure 5.7d).[3,12]

(a)   (b)

(c)   (d)

FIGURE 5.7   Cutting process: (a) placing cuttings in the containers filled with a mixture of peat and perlite; (b) high humidity content in the environment of the rooting greenhouse; (c) already rooted cuttings before transplanting; (d) just transplanted rooted cuttings. (Courtesy of Cuenca, B.)

FIGURE 5.8   The C-box cutting system developed by the Chestnut R&D Center Piemonte. (Courtesy of Beccaro, G.)

The rooting system of a plant developed from a cutting is usually better and more balanced than that of a stooled plant. However, cutting development can be limited by the pot size. Cuttings transplanted to a 2 L pot can be grafted the following year. Micrografting onto a one-year cutting is also possible.

5.3.3 Micropropagation

Micropropagation, using *in vitro* tissue culture, is an alternative to the traditional vegetative propagation methods. Classic micropropagation involves the development of microcuttings in a semisolid aseptic culture media, confined in closed vessels. Such plants are heterotrophic and photosynthetically inactive, with dysfunctional stomata that do not allow gas exchange. Culture media, therefore, should contain a source of carbohydrates as well as mineral salts, vitamins, growth regulators and agar, if semisolid culture is performed.

This method presents different opportunities, since micropropagated plants produced more vigorous growth in the field than plants propagated by other methods, due to their profuse and well-balanced root system. It is an ideal technique for the rapid propagation of clonal rootstocks or self-rooted trees year-round, starting with little initial material. The production of plants free from ChMV, transmitted by *Myzocallis castanicola*, or any other virus or disease is another benefit of micropropagation, as it is an aseptic, controlled process. Problems related to the loss of rooting ability of adult genotypes can also be easily overcome as micropropagation can use ontogenically young materials such as epicormic buds, etiolated or basal sprouts, spheroblasts[13] that allow adult genotypes to be cloned. Moreover, introducing changes in the genotypes that a nursery is producing is quick and easy. However, the need for specialized facilities, the high demand for skilled labor, and the acclimatization requirement makes this method more expensive than traditional ones.

Micropropagation protocols have been developed by several different research teams.[8,14–25] Micropropagation starts by establishing material of a certain genotype in *in vitro* culture. Juvenile, dormant plant winter material is collected, disinfected, and forced to sprout in a controlled environment. The new sprouts are surface-sterilized in a solution of hypochlorite and established in culture. After subcultures are stabilized by eliminating contaminated or non-reactive explants, commercial propagation begins. The process is carried out in several phases: a multiplication phase, in which new shoots are generated from the initial one; an elongation phase, in which the generation of new shoots stops and the existing ones grow to a suitable size for rooting; a rooting phase, in which the basal portion of elongated shoots are dipped in a sterile concentrated auxin solution (1 g/L) for rooting; and the acclimatization phase (hardening), in which the rooted explants switch from a heterotrophic to an autotrophic habit. Acclimatization is difficult without a controlled environment because explants are very sensitive to dehydration.

In classic micropropagation, all of these phases are carried out in semisolid media, using those described by Gresshoff and Doy (1972) and modifications of Murashige and Skoog (1962). However, roots produced *in vitro* frequently have low functionality with poor uptake of water and nutrients, and plants fail to, resulting in plant loss during the acclimatization process (Figure 5.9a–d).

(a)   (b)

(c)   (d)

FIGURE 5.9 Acclimatization Micropropagation process: (a) field-collected chestnut cuttings chestnut from a resprouted stump and forced to grow in a controlled chamber; (b) multiplication phase in classic semisolid culture; (c) elongation phase in classic semisolid culture; (d) elongation phase in temporary immersion system (TIS). (Courtesy of Cuenca, B.)

5.3.3.1 Temporary Immersion Systems

To overcome plant loss during micropropagation, a temporary sterile immersion systems (TIS culture) is used during the elongation phase. These systems use a special vessel that contains the liquid media and that pumps the sterile air into the vessel through a 0.2 μm filter. The incoming air has a double function: it makes the liquid move upward to the plants to feed them or back down to a reservoir, and it renews the atmosphere inside the container, allowing gas exchange to occur via stomata to keep the explant functional (Figure 5.9d).[24,28] The elongated shoots generated this way are then rooted *ex vitro* and acclimatized simultaneously. Growth chambers or temperature-controlled greenhouses, which maintain high humidity and a temperature around 25°C, are essential for success. More recently, Spanish researchers have developed a photoautotrophic micropropagation (PAM) system for chestnut culture,[29,30] in which a temporary or continuous immersion system is used under photosynthetically active radiation (PAR) lights and the incoming air is supplemented with CO_2 (Figure 5.10a and b).

(a) (b)

FIGURE 5.10 Photoautotrophic micropropagation: (a) PAM rooted shoots before acclimatization; (b) acclimatized vitroplants ready for nursery growth. (Courtesy of Cuenca, B.)

PAR lights emit in the same spectral range as solar radiation (400 to 700 nm) that plants use for photosynthesis. Moreover, CO_2 supplied at 2,000–3,000 ppm allows the plantlets in culture to be photosynthetically active so sucrose in the media can be suppressed. The entire micropropagation process can be using supplemental lighting and CO_2 in this way, but the effect is especially remarkable during the rooting process (Figure 5.11a and b).

(a) (b)

FIGURE 5.11 Rooting process: (a) PAM rooting of TIS elongated shoots; (b) *ex vitro* rooting/acclimatization of TIS elongated shoots. (Courtesy of Cuenca, B.)

Explants produced by the TIS and PAM system and placed in a rockwool block have high rooting rates, and during acclimatization stress is eliminated, since the explants are already autotrophic. The roots developed appear normal and functional. Acclimatized *in vitro* plants are normally transplanted into large pots during the spring and grown until they reach the size for grafting. Large-scale chestnut micropropagation is feasible with an investment in laboratory and acclimatization facilities

5.4   GRAFTING AND BUDDING TECHNIQUES

Although grafting still presents some challenges due to the potential for rootstock/scion incompatibility (especially for clonal rootstocks) and/or canker blight infection, it remains the most common chestnut propagation method used by nurseries. Factors needed for grafting success include suitable propagation material, the correct technique at the proper time, grafter skill, time and optimal environmental conditions (mild temperatures with little wind and moisture control), and rootstock/scion genetic compatibility.

A list of the main grafting and budding techniques is reported in Table 5.1.

5.4.1   Scion: Harvest and Cold Storage

Harvest and cold storage of scion wood is important for successful grafting. Ideally, scion wood is collected from genetically certified, true-to-type mother trees. The scions are cut from healthy and well-lignified one-year-old branches of mother trees, which were prepared during the previous year by severe pruning. When the mother plant is an old specimen, severe pruning is needed to generate new shoot growth suitable for scion wood. Scion wood is cut when trees are fully dormant in

---

TABLE 5.1

Grafting and Budding Techniques and Timing

Grafting	Timing
Triangle (inlay) graft	Winter–spring
Splice (slice) graft	Winter–spring
Cleft graft	Early spring
Whip graft	Winter
Lateral graft	Early spring
Under bark graft	Spring
Crown graft	Spring
Chip budding	Summer
T-budding	Spring–summer
Cadillac	Early spring
Semi soft scion graft	Summer

Source: Bounous, G., *Il castagno: Risorsa multifunzionale in Italia e nel mondo*, Edagricole, Bologna, Italy, p. 420, 2014.

---

mid-winter and is cold-stored, slightly moistened, inside polythene bags, at +1°C, +2°C. In regions where Asian gall wasp is problematic, hot water immersion of chestnut scion wood is used to induce larval mortality in infested buds. After the immersion of scion wood in hot water at 52°C for 10 minutes, it is air dried for 30 minutes, sealed in polyethylene bags, and cold-stored for later grafting.

5.4.2 GRAFTING COMPATIBILITY AND TIME OF GRAFTING AND BUDDING

C. mollissima is usually grafted onto its own seedlings. To graft European cultivars, it is better to use *C. sativa* or Eurojapanese hybrid seedlings or clonal rootstocks. However, for a *C. crenata* × *C. sativa* hybrid or a *C. crenata* cultivar, it is better to use a *C. crenata* or a hybrid seedling rootstock. Incompatibility often occurs between European and Asian species. The main period for grafting is from late winter to early spring, and from early summer (micrografting) to mid-to-late summer (budding), depending on the type of graft and the local climate. In the USA, nurseries use seedling-grown rootstocks for grafting in a greenhouse. For graft union success, moisture content of the growing medium is important. Medium moisture content near 48% results in greater grafting success than when the moisture is above 56%. Also, graft unions are covered with aluminum foil, maintained in a greenhouseat 24°C day/18°C night on a 12 h cycle with minimal irrigation four weeks before placing them outdoors under 55% shade cloth.

5.4.3 WINTER AND SPRING GRAFTS

Triangle (inlay) grafting: The triangle graft can be used in the nursery or in an orchard on rootstocks 2–3 cm in diameter. It consists of creating a triangular wedge, which is first made in the rootstock. A similar wedge-shaped cut is made on the basal end of the dormant scion before inserting it into the cut portion of the rootstock with a scion bud oriented outward[31,32] (Technical Sheet 5.2). Contact between the rootstock and scion cambium is essential. A dressing is applied at the union and on the cut surface of the scion apex (Figure 5.12a and b).

(a)   (b)

FIGURE 5.12 Triangle graft: (a) triangular wedge is cut in the rootstock; (b) scion inserted in the rootstock. (Courtesy of Beccaro, G.)

TECHNICAL SHEET 5.2   TRIANGLE (INLAY) GRAFT

Execution: end of the winter and spring

1. A triangular wedge is created in the rootstock.

2. The scion is shaped.

3. The scion is inserted in the rootstock and dressing is applied.

Courtesy of Facello, V.

FIGURE 5.13   Splice graft. (Courtesy of Beccaro, G.)

Splice (slice) graft: Currently used in nurseries and in the field when the scion and the rootstock have the same diameter (Figure 5.13); the scion is joined onto the stem of a rootstock. The junction must be secured with a rubber grafting strip or twine and sealed with dressing.

Cleft graft: This simple form of grafting is performed when the rootstock has a larger diameter than the scion. The rootstock first topped and then split to accommodate the insertion of two scions. A long, tapering wedge is made at the basal end of each of two scions before inserting them at the outer edge of the split in the rootstock. The cambium of the scion and rootstock must be properly aligned for graft union success (Technical Sheet 5.3).

TECHNICAL SHEET 5.3   CLEFT GRAFT

Execution: early spring

1. The rootstock is cleft.

2. Two scions are inserted in the cleft, one at each side. Dressing is applied.

Courtesy of Facello, V.

Whip graft: This technique, also called whip and tongue graft (Technical Sheet 5.4), is preferred to cleft graft because scion and rootstock can be securely positioned with their cambial tissue aligned and tightly wrapped to promote callus formation and successive graft union success (Figure 5.14). For this graft, the scion and rootstock diameters should be similar. Also, the

FIGURE 5.14   Whip graft. (Courtesy of Beccaro, G.)

cuts made on the rootstock should match those on the scion. For both the rootstock and scion, a sloping cut is made using a single knife stroke. Next, a reverse cut is made to form the "tongue", starting one-third of the distance from the tip to about one-half the length of the first cut. The scion is then inserted into the rootstock, interlocking the tongues of both pieces. As with other grafts, alignment of rootstock and scion cambial tissues is essential.

TECHNICAL SHEET 5.4   WHIP GRAFT

Execution: end of winter, early spring

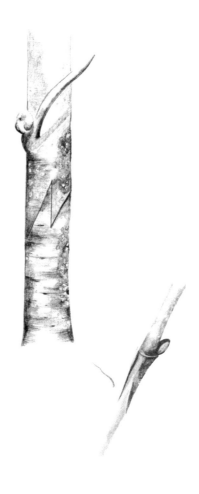

1. The interlocking tongues shaped in the rootstock and the scion must have the same size.

2. The graft is wrapped tightly: a dressing is applied only on the top of the scion. Alternatively, the union and the top of the scion can be sealed with parafilm, before covering the union with aluminum foil.

Courtesy of Facello, V.

(a)   (b)

FIGURE 5.15   Under bark grafting (a) with one scion disposed under the bark (b). (Courtesy of Beccaro, G.)

Lateral graft: This type of graft is made in early spring, but is difficult to perform. When the callus is formed, the rootstock is pruned heavily to promote the sprout of the buds from the scion.

Under bark graft: This is performed in the spring at the beginning of vegetative growth when the bark slips easily (Technical Sheet 5.5). It is preferable when the diameter of the rootstock is too large to be used for cleft or whip grafting. However, the under bark graft can produce a weak union, which can break under windy conditions due to the position of the scion into the rootstock. For this graft, the rootstock is first topped. Then, only one cut along one side of the basal portion of the scion is made, leaving a "shoulder" at the upper part of the cut. The scion is inserted underneath the rootstock bark with the scion shoulder placed on top of the rootstock stub. The graft is then secured and a dressing is applied (Figure 5.15a and b).

TECHNICAL SHEET 5.5   UNDER BARK GRAFT

Execution: early to late spring

1. A vertical slit is performed through the bark where the scion will be inserted.

2. The scion is placed behind the flap of
bark and wrapped tightly. Dressing
must be applied on the wounds.

More than one scion can be inserted
in the rootstock (crown graft).

Courtesy of Facello, V.

5.4.4 SUMMER GRAFTS

Chip budding: For this technique, a small section of scion wood with a bud
(i.e., a "chip") is placed onto the rootstock where a similar shaped wedge cut
has been made (Technical Sheet 5.6). Chip grafting is done in early cold-
stored scion wood. When chip-budding, the irregular shape of the vascular
cambium can make the alignment of scion and rootstock cambium difficult. The non-uniform depth of the cambium at various positions around
the stem often results in misalignment of scion tissue with rootstock tissue,
preventing the formation of a callus bridge and subsequent establishment
of vascular continuity necessary for graft union success. Chip budding can
also be performed in late summer with leaves removed from the buds before
grafting (Figure 5.16a–c). In this case, growth from the scion bud begins the
following spring. After grafting or budding, any shoots developed below the
graft union must be removed. When growth from more than one scion bud
occurs, the weaker shoot is eliminated. Also, the new scion growth is often
supported with a stake to ensure a straight tree.

TECHNICAL SHEET 5.6   CHIP BUDDING
Execution: early to late spring

1. A scion bud is placed on a rootstock where angled cuts are made to exactly correspond with those on the bud chip.

2. The cambium layers should match as exactly as possible.

3. The graft is bound tightly with tape. Once growth occurs from the bud, the rootstock tissue above the graft is removed and a dressing is applied to cut.

Courtesy of Facello, V.

(a)   (b)

(c)

FIGURE 5.16 Chip budding: (a) two cuts in the rootstock are performed and the resulting sliver of is removed; (b) bud chip placed on the cut; (c) the joint is bond tightly with grafting tape or strips of polythene. (Courtesy of Beccaro, G.)

> *T-budding*: This is a common spring or summer graft, which is used when the bark easily separates from the wood. For this graft, a T-shaped cut is made on the rootstock. Next, a similar-sized shield with a single bud is cut from the scion and is then inserted under the flaps of the rootstock bark (Technical Sheet 5.7).

TECHNICAL SHEET 5.7   T-BUDDING

Execution: late spring to end of the summer

1. Performed when the bark easily lifts in one uniform layer from the wood.

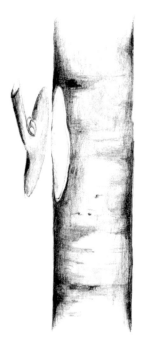

2. The bud and a small sliver of the wood underneath it are cut off from the scion.

3. The bud shield is placed into the T flaps of the stock.

4. The bud shield is wrapped with a tape. In the spring, when the bud begins growth, the upper part of the rootstock tissue above the bud union scion is cut off and a dressing is applied on the cut.

Courtesy of Facello, V.

Soft and in vitro cleft grafts: Semi-soft scion grafts are used on rootstocks grown in pots during the growth period. Grafted plants are forced in the greenhouse under mist and the graft union can be held in place with plastic clips. This type of graft is commonly done on *in vitro* plants, and it has some advantages. First, the physiological stage of the tissues facilitates the healing of the graft union, so callus formation occurs within a month and the rate of success is high when compared to traditional systems. Second, as the rootstock tissue is young and soft, if the graft fails, it can be pruned and re-grafted, using a traditional method the next season, with no apparent deformations.[25]

5.4.5   Epicotyl Grafting

The epicotyl grafting technique has been developed on *C. mollissima*. Seeds are harvested from September to October (northern hemisphere), air-dried for 7–10 days, packed in bags, and then stored at 4°C for stratification. In January, seeds are soaked in running cold water for 1–2 hours and then placed in moist sand until May. At this time seedlings are used as that have a 2–3 mm diameter epicotyl. Seedlings are dug from the nursery and washed to remove the sand. Actively-growing, 15-cm-long, 1-year-old scions of chestnut are collected with at least four nodes on each scion stick (2–3 mm diameter) to match the diameter of seedling epicotyl.

For grafting, the rootstock is cut off smoothly above the epicotyl. Next, the center of the epicotyl is split 2 cm deep with a grafting knife (Figure 5.17a and b). Then sloping cuts are made on both sides of the scion, 2 cm from the lower end (Figure 5.17c) before it is fitted and inserted on the rootstock (Figure 5.17d). The cut area and the scion are wrapped with grafting tape to seal the cuts and scion (Figure 5.17e). The completed graft (Figure 5.17f and g) is transferred to a 12 × 15 cm container with one or two buds out of the substrate.[33,34] The buds of scion will grow after 40 days (Figure 5.17h). The success of this technique is about 70%, but is not currently used by commercial nurseries.

5.4.6   Inverted Radicle Grafting

In inverted radicle grafting, germinated seeds with radicles 6–7 cm long are used as rootstocks (Figure 5.18a). The radicle tip is cut off and split. Next, the scion (4–5 cm long and 4–6 mm in diameter) is prepared with a wedge-shaped cut (Figure 5.18b). Rootstock and scion are cleft grafted and, afterward, the graft is wrapped tightly with a parafilm strip 3–4 mm wide (Figure 5.18c), and the scion apex is protected with a dressing (Figure 5.18d). The graft success and final survival is about 70%–80%.

5.4.7   Graft Care

Graft wound protection prevents tissue dehydration and pathogen infection. Hot or cold commercial dressings are often used, which are composed of fats, natural and synthetic resins, or vegetable oils. A good dressing is elastic and impermeable, and contains compounds that prevent or reduce *Cryphonectria parasitica* infection, especially in the early years of grafting. To protect wounds, formulations containing hypovirulent strains of *C. parasitica* can be applied to grafts. Formulations containing copper or benzimidazole fungicides mixed with mineral oil provide good protection against the fungus.

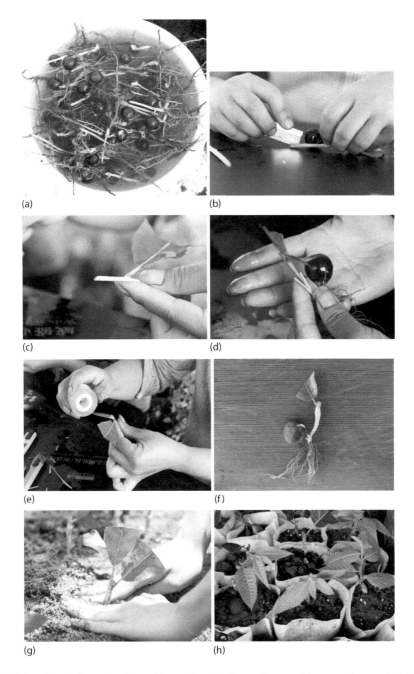

FIGURE 5.17 Epicotyl grafting: (a) seedlings (2–3 mm-diameter) harvested around 10 May; (b) bisected the epicotyl; (c) sloping cuts made on both sides of the scion; (d) the scion inserted into the epicotyl; (e) the graft and the scion wrapped with grafting tape; (f) a completed graft; (g) new graft in the nursery; (h) 40 days after grafting in the nursery. (Courtesy of Zou, F.)

(a)   (b)

(c)   (d)

FIGURE 5.18 Inverted radicle grafting: (a) radicles 6–7 cm long are used as rootstocks with the radicle tip removed; (b) the radicle is split and the scion is prepared with a wedge shaped cut; the scion is inserted into the radicle and the graft is wrapped tightly with a strip of parafilm; (d) grafted seed in pot. (Courtesy of Bounous, G.)

In the spring, shoot growth occurs rapidly on the scion. The strongest shoot is selected and supported to encourage straight growth. Non-biodegradable grafting tape (or parafilm) should be removed it restricts growth (Figure 5.19). Young grafted trees also require fertilization, fungicide and insecticide treatments, as well as irrigation.

FIGURE 5.19   Post grafting dressing. (Courtesy of Beccaro, G.)

5.4.8 INCOMPATIBILITY CAUSES AND SYMPTOMS

Since new interspecific hybrid chestnut cultivars and rootstocks were developed, low graft success and graft incompatibility become problematic. Early incompatibility may occur within two years of grafting and late incompatibility can develop in 5 to 7 years.[35]

Genetic causes

- Grafting between genus (e.g. onto oak)[36–38];
- Grafting between species;
- Grafting with hybrids;
- Grafting in species (in some Chinese and European species).[39–42]

Anatomic causes

- The 5-lobed ring of cambial tissue causes poorly matched or misaligned scion and rootstock tissues[43,44] and irregular shaped cambium tissue;
- Interruptions in cambial connection;
- Thick necrotic layer between scion and rootstock.

Biochemical causes

- Presence of different cambial isoperoxidase isozymes[45];
- Potential effect of different types of phenolic compounds like presence of prunasine and similar inhibitory substances related with incompatibility.

Biotic causes: Infection with virus or mycoplasma (Chestnut Mosaic Virus).[46]
Not all graft failures are a result of graft incompatibility. Graft failure can be
  also caused by low temperature injury during winter, other adverse environmental conditions, infection of the graft union by chestnut blight, or
  improper grafting techniques.

Symptoms

- Very low graft success;
- Absence of a normal cambium tissue in the graft area and the lack of vascular continuity between the rootstock and scion (Figure 5.20a);

(a)

(b)   (c)

FIGURE 5.20 Graft incompatibility symptoms: (a) the absence of a normal cambium tissue in
the graft area and the lack of vascular continuity between the rootstock and scion; (b) yellowing
of the leaves during the growing season, followed by early premature defoliation; (c) scion
desiccation. (Courtesy of Serdar Ü.)

- Yellowing of the leaves during the growth period, followed by early foliage (Figure 5.20b);
- Shoot decline, foliar chlorosis, and small leaves (see Chapter 13);
- Scion desiccation (Figure 5.20c);
- Overgrowth at the graft union (see Chapter 13);
- Excessive thickness of tissue around the scion bud (Figure 5.21a);
- Presence of a necrotic layer between the graft components and, consequently, breakage at the graft (Figure 5.21a);
- Excessive suckering on the rootstock (Figure 5.21b);

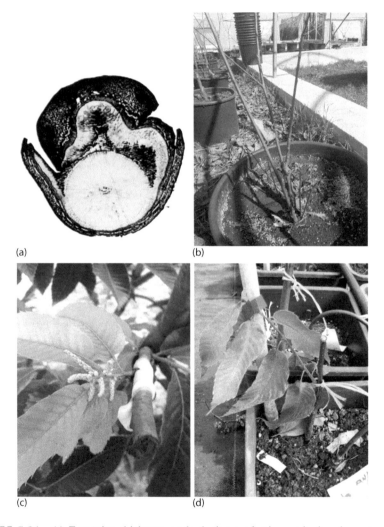

(a)   (b)

(c)   (d)

FIGURE 5.21   (a) Extensive thickness on bark tissue of scion or bud and presence of a necrotic layer between the graft components and, consequently, breakage at the graft; (b) excessive rootstock suckering; (c) scion shoots become stunted and produce flowers; (d) rootstock and scion mortalities die with scion. (Courtesy of Serdar Ü.)

- Scion shoots become stunted and produce flowers (Figure 5.21c);
- Rootstock and scion mortality (Figure 5.21d).

Except for the last symptom, one or more of these symptoms may not necessarily mean scion/rootstock the combination is incompatible.

REFERENCES

1. Conedera, M., Krebs, P., Tinner, W., Pradella, M. and Torriani, D. 2004a. The cultivation of *Castanea sativa* (Mill.) in Europe, from its origin to its diffusion on a continental scale. *Vegetation History and Archaeobotany* 13(3): 161–179.
2. Cuenca, B. and Majada, J. 2012. *Castanea sativa* Mill. In: García, J.P., Cerrillo, R.M.N., Peragón, J.L.N., Prada Sáez, M.A., Serrada Hierro, R. Producción y manejo de semillas y plantas forestales. *Tomos I y II. Organismo autónomo Parques Nacionales. Ministerio de Agricultura, Alimentación y Medio Ambiente.*
3. Bounous, G. 2014. *Il castagno: Risorsa multifunzionale in Italia e nel mondo.* Edagricole, Bologna, p. 420.
4. Fernandez, J., Pereira, S. and Miranda, E. 1992. Fog and substrate conditions for chestnut propagation by leafy cuttings. *Proceedings publiés par AFOCEL, Production de varietes genetiquement ameliorees d'especes forestieres a croissance rapide, Mass Production Technology for Genetically Improved Fast Growing Forest Tree Species, Bourdeaux* 379–383.
5. Bounous, G. and Marinoni, D.T. 2005. Chestnut: Botany, horticulture, and utilization. *Horticultural Reviews* 31: 291–347.
6. Bonga, J. and Durzan, D. 1982. *Tissue Culture in Forestry.* Springer Science & Business Media, Dordrecht, the Netherlands.
7. Mellano, M.G., Saggese V. and Donno, D. 2018. Substrati di coltivazione per l'allevamento di portinnesti clonali. Growing medium for *Castanea* clonal rootstocks. *Castanea* 11: 14–15.
8. Cuenca, B., Rodríguez, L., Cámara, M. and Ocaña, L. 2005. Micropropagación de ejemplares adultos de *Castanea sativa* Mill. seleccionados por resistencia a *Phytophthora cinnamomi. Actas de la VI Reunión de la Sociedad Española de Cultivo In Vitro de Tejidos Vegetales* 11–13.
9. Ocaña, L., Santos, M., Gómez, J. and Cu, B. 2001. Reproducción en vivero de castaños híbridos resistentes a la tinta mediante estaquillado semiherbáceo, *Congresos Forestales.*
10. Rodriguez, L., Cuenca, B. and Pato, B. 2005. Producción in vitro y mediante estaquillado semiherbáceo de híbridos de castaño en un vivero comercial: Coste y eficiencia de ambos sistemas, *Congresos Forestales.*
11. Beccaro, G., Alma, A., Gonthier, P., Mellano, M., Ferracini, C., Giordano, L., Lione, G., Donno, D., Boni, I. and Ebone, A. 2017. Chestnut R&D Centre, Piemonte (Italy): 10 years of activity. *Acta Horticulturae* 1220: 133–140.
12. Donno, D., Beccaro, G.L., Mellano, M.G., Bonvegna, L. and Bounous, G. 2014. *Castanea* spp. buds as a phytochemical source for herbal preparations: Botanical fingerprint for nutraceutical identification and functional food standardisation. *Journal of the Science of Food and Agriculture* 94(14): 2863–2873.
13. Sanchez, M., Carmen San-Jose, M., Ferro, E., Ballester, A. and Vieitez, A.M. 1997. Improving micropropagation conditions for adult-phase shoots of chestnut. *Journal of Horticultural Science* 72(3): 433–443.
14. Vieitez, A. and Vieitez, M. 1983. *Castanea sativa* plantlets proliferated from axillary buds cultivated *in vitro. Scientia Horticulturae* 18(4): 343–351.

15. San José, M.C. and Vieitez, E. 1984. Regeneración in vitro de plantas de castaño a partir de yemas adventicias. In: *Congreso sobre el castaño, Lourizán* 391–395.
16. Vieitez, A.M., Sánchez, C. and San-José, C. 1989. Prevention of shoot-tip necrosis in shoot cultures of chestnut and oak. *Scientia Horticulturae* 41(1–2): 151–159.
17. Sánchez, M. and Vieitez, A. 1991. In vitro morphogenetic competence of basal sprouts and crown branches of mature chestnut. *Tree Physiology* 8(1): 59–70.
18. Ballester, A., Sanchez, M.C. and Vieitez, A.M. 1992. New strategies for in vitro propagation of chestnut. In: *Proceedings of the Wild Chestnut Industry Conference*, Morgantown, USA, 32–40.
19. Miranda, M.E. and Fernandez, J. 1992. Micropropagation as a nursery technique for chestnut hybrid clones. *Proceedings of the International Chestnut Conference*, West Virginia University Press, Morgantown, VA, 101–103.
20. Seabra, R. and Pais, M. 1998. Genetic transformation of European chestnut. *Plant Cell Reports* 17(3): 177–182.
21. Gonçalves, J.C., Diogo, G. and Amâncio, S. 1998. In vitro propagation of chestnut (*Castanea sativa* × *C. crenata*): Effects of rooting treatments on plant survival, peroxidase activity and anatomical changes during adventitious root formation. *Scientia Horticulturae* 72(3–4): 265–275.
22. Cuenca, B., Ocaña, L., Salinero, C., Pintos, C., Mansilla, J. and Rial, C. 2009. Selection of *Castanea sativa* Mill. for resistance to *Phytophthora cinnamomi*: Micropropagation and testing of selected clones. In: *European Congress on Chestnut-Castanea 2009*, 111–119.
23. Oakes, A.D., Powell, W.A. and Maynard, C.A. 2013. Doubling acclimatization survival of micropropagated American chestnuts with darkness and shortened rooting induction time. *Journal of Environmental Horticulture* 31(2): 77–83.
24. Vidal, N., Blanco, B. and Cuenca, B. 2015. A temporary immersion system for micropropagation of axillary shoots of hybrid chestnut. *Plant Cell, Tissue and Organ Culture (PCTOC)* 123(2): 229–243.
25. Cuenca, B., Lario, F., Luquero, L., Ocaña, L. and Mandujano, M. 2017. Early grafting of chestnut by green grafting. *Acta Horticulturae* 1220: 141–148.
26. Gresshoff, P.M. and Doy, C.H. 1972. Development and differentiation of haploid *Lycopersicon esculentum* (tomato). *Planta* 107(2): 161–170.
27. Murashige, T. and Skoog, F. 1962. A revised medium for rapid growth and bio assays with tobacco tissue cultures. *Physiologia Plantarum* 15(3): 473–497.
28. Vidal, N., Correa, B., Rial, E., Regueira, M., Sánchez, C. and Cuenca, B. 2015. Comparison of temporary and continuous immersion systems for micropropagation of axillary shoots of chestnut and willow. *Acta Horticulturae* 1083: 227–233.
29. Cuenca, B., Sánchez, C., Aldrey, A., Bogo, B., Blanco, B., Correa, B. and Vidal, N. 2017. Micropropagation of axillary shoots of hybrid chestnut (*Castanea sativa* × *C. crenata*) in liquid medium in a continuous immersion system. *Plant Cell, Tissue and Organ Culture (PCTOC)* 131(2): 307–320.
30. Aldrey, A., Blanco, B., Bogo, B., Cuenca, B., Sánchez, C., Luquero, L., Ocaña, L., Mandujano, M. and Vidal, N. 2018. Photomixotropic and photoautotrophic micropropagation of *Phytophthora* resistant chestnut genotypes using liquid media. *Acta Horticulturae* 1220: 177–184.
31. Mellano, M.G., Donno, D. and Beccaro, G.L. 2015. Innovazioni nel vivaismo: il microinnesto di *C. sativa* e ibridi.
32. Mellano, M.G., Cerutti, A.K. and Beccaro, G.L. 2016. Castanicoltura ad alta densità: aspetti di sostenibilità. *Castanea* 5: 8–9.
33. Zhang, R., Peng, F.-R., Le, D.-L., Liu, Z.-Z., He, H.-Y., Liang, Y.-W., Tan, P.-P., Hao, M.-Z. and Li, Y.-R. 2015. Evaluation of epicotyl grafting on 25-to 55-day-old Pecan seedlings. *HortTechnology* 25(3): 392–396.

34. Tan, L.-M., Yuan, D.-Y., Zhang, D.-Q., Xiong, H., Liu, D.-M., Zhang, X.-H. and Zhu, Z.-J. 2016. Grafting techniques in *Castanea henryi* using bud seedlings as rootstocks. *Nonwood Forest Research* 34(3): 153–157.

35. Oraguzie, N., McNeil, D. and Thomas, M. 1998. Examination of graft failure in New Zealand chestnut (*Castanea* spp) selections. *Scientia Horticulturae* 76(1–2): 89–103.

36. Ada, S. and Ertan, E. 2013. Histo-cytological study of the graft union of the chestnut/ Oak. *Agriculture, Forestry and Fisheries* 2(2): 110–115.

37. Heitz, R. and Jacquiot, C. 1972. Étude anatomique de la greffe d'un châtaignier sur chêne. *Annales des Sciences Forestières*, 391–395.

38. Park, K.S. 1968. Studies on the juvenile tissue grafting of some special use trees III. On the modified nurse seed grafting of some crop tree species (chestnut, ginkgo and oak). *Research. Institute On Fororestry and Genetics Korea* 6: 89–104.

39. Mac Daniels, L. 1955. Stock-scion incompatibility in nut trees. *Annual Report of the Northern Nut Growers Association* 46: 92–97.

40. Santamour, F.S. 1988. Graft incompatibility related to cambial peroxidase isozymes in Chinese chestnut. *Journal of Environmental Horticulture* 6(2): 33–39.

41. Serdar, U., Demirsoy, H., Macit, I. and Ertürk, U. 2010. Graft compatibility in some Turkish chestnut genotypes (*C. sativa* Mill.). *Acta Horticulturae* 866: 285–290.

42. Serdar, U. and Soylu, A. 2005. The effect of grafting time and methods on chestnut nursery tree production. *Acta Horticulturae* 693: 187–194.

43. Warmund, M.R., Cumbie, B.G. and Coggeshall, M.V. 2012. Stem anatomy and grafting success of Chinese chestnut scions on 'AU-Cropper' and 'Qing' seedling rootstocks. *HortScience* 47(7): 893–895.

44. Huang, H., Norton, J., Boyhan, G. and Abrahams, B. 1994. Graft compatibility among chestnut (*Castanea*) species. *Journal of the American Society for Horticultural Science* 119(6): 1127–1132.

45. Santamour, F., McArdle, A. and Jaynes, R. 1986. Cambial isoperoxidase patterns in *Castanea*. *Journal of Environmental Horticulture* 4(1): 14–16.

46. Desvignes, J. 1999. Sweet chestnut incompatibility and mosaics caused by the chestnut mosaic virus (ChMV). *Acta Horticulturae* 494: 451–458.

6 Orchard Management

Gabriele Beccaro, Giancarlo Bounous, José Gomes-Laranjo, Michele Warmund, and Jane Casey

CONTENTS

6.1   ORCHARD ESTABLISHMENT

6.1.1   SELECTING THE SITE

Establishing chestnut orchards with different *Castanea* species should be carried out with consideration for climate, soil, altitude, rainfall, and other parameters to ensure good, high quality productions.

6.1.1.1   Soil

The best soils for chestnut are deep, soft, volcanic, and rich in phosphorus and potassium (Table 6.1). The pH should range from 5.0 to 6.5. Soil with active limestone must be avoided, because *Castanea* is very sensitive to high pH. Chestnut prefers soil with 2%–3% or more organic matter. Soil permeability is very important. Trees perform best in well-drained loam to sandy-loam soils. Heavy, washed out, clayey, stagnant soils favor root rot caused by *Phytophthora* spp. and *Armillaria mellea* and must be avoided.

TABLE 6.1

Soil Analysis Benchmark for Chestnut

Elements	Low Level	Medium Level	High Level
Total Nitrogen (‰)			
Sandy soil	<0.8	0.8–1.2	>1.2
Loam soil	<1.0	1.0–1.6	>1.6
Clay soil	<1.2	1.2–1.6	>1.6
Organic Matter (%)			
Sandy soil	<0.8	0.8–1.5	>1.5
Loam soil	<1.5	1.5–2.0	>2.0
Clay soil	<2.0	2.0–2.5	>2.5
Assimilable phosphorus (ppm) (Olsen Method)	<10	10–20	>20
Cation-exchange capacity (C.E.C) (meq/100g)	<10	10–20	>20
Exchangeable Potassium (ppm) with C.E.C			
<10	<70	70–120	>120
10–20	<100	100–200	>200
>20	<150	150–300	>300
Exchangeable Calcium (ppm) with C.E.C			
<10	<800	800–1800	>1800
10–20	<1500	1500–3500	>3500
>20	<3000	3000–6000	>6000
Exchangeable Magnesium (ppm) with C.E.C			
<10	<70	70–120	>120
10–20	<100	100–180	>180
>20	<150	150–300	>300

Source: Bounous, G. *Il castagno: Risorsa multifunzionale in Italia e nel mondo.* Edagricole, Bologna, Italy, p. 420, 2014.[13]

6.1.1.2   Climate

Chestnut tolerates cold winters and is suitable for an average temperature of 8°C–15°C, with an average of 10°C per month for at least 6 months. *C. sativa* is more cold-resistant (−15°C—20°C) than many of the Eurojapanese hybrids. In spite of late bud-break (March–April), the trees may be subject to spring frosts which damage tender growing shoots. During blossoming and pollination, temperatures of 27°C–30°C are necessary. European and Japanese cultivars require about 800–900 mm/year of rainfall, well distributed during the growing season. Eurojapanese hybrids and *C. mollissima* have a higher water requirement (1200–1300 mm/year). Water availability for irrigation should be considered. In temperate climates, *C. sativa* should not be planted above 700–800 m, while for hybrids the maximum altitude is about 500–600 m.

6.1.2   How Many Trees?

For all species, except in traditional European chestnut orchards, the general trend for orcharding is to increase tree density to develop maximum bearing per unit area in a relatively short time. Plantation densities can range from 100 to 550 trees/ha, based on species, genotype-environment interactions, and cultural practices.[1] For traditional plantations of *C. sativa*, spacing ranges from 8–10 m apart in rows and 8–12 m between rows (Figure 6.1a and b). For the most vigorous Eurojapanese hybrids the distances range between 7 × 8 m (178 trees/ha) and 8 × 10 m (125 trees/ha). Both *C. sativa* and Eurojapanese hybrids can be cultivated at high densities (3 m × 10 m). For *C. crenata*, spacings of about 5 × 7 m (285 trees/ha in fertile soils) or 7 × 7 m (204 trees/ha) are recommended. *C. mollissima*, both in the USA and China, is usually grown at a high/semi-high density, due to the smaller tree size. Planting patterns may be square, rectangular, or triangular, but rectangular and square are the most common because they are easier to manage.

Orchard architecture depends strongly on species and country (see Chapters 7 and 8). In Australia, for example, the early orchard layouts (*C. sativa* and Eurojapanese hybrids) were 13 × 13 m, but new orchards are now generally planted at 7 × 7 m (diamond layout), with some intensively planted orchards at 3.5 × 7 m (square layout). The aim of intensively planted orchards is to gain early production and then cull every second tree after 8–9 years.

6.1.3   Rootstocks

Seedlings or clonal rootstocks can be used (see Chapter 5). Clonal rootstocks are not available for *C. mollissima*. However, European and Japanese chestnut can be grafted onto interspecific hybrids selected in Portugal, Spain, and France that have resistance to ink disease, easy propagation, and good compatibility. The most popular clonal rootstocks are still the Eurojapanese hybrids selected in France by INRA in the second half of the twentieth century. These hybrids are easy to propagate by layering or soft cuttings. The INRA hybrid rootstocks are also tolerant to *Phytophthora* spp., but are somewhat sensitive to *Cryphonectria parasitica*, and have genetic compatibility with most *C. sativa* cultivars. Popular rootstocks include: CA 07 'Marsol' (moderately resistant to *Phytophthora*); CA 74 'Maraval' (*Phytophthora* resistant, low vigour); CA 118 'Marlhac' (moderately resistant to *Phytophthora*, able to grow at temperatures <−10°C); CA 90 'Ferosacre' (*Phytophthora* resistant, but sensitive to temperatures <−10°C). The Portuguese selection CoIUTAD is *Phytophora* and *Dryocosmus kuriphilus* resistant. European chestnut cultivars are usually grafted onto seedlings of *C. sativa*, as shown in Table 6.2.[2,3] Resistance to ink disease (*Phytophthora spp.*) in Spanish hybrids varies from medium to very resistant.[4]

(a)

(b)

FIGURE 6.1   (a) Intensive Eurojapanese hybrids plantation with 6 × 3.5 m rows, Northern Italy; (b) intensive 15 years Eurojapanese hybrids plantation with 6 × 6 m rows, Northern Italy.

(Continued)

FIGURE 6.1 (Continued) (c) *C. mollissima* intensive orchard in China; (d) *C. sativa* high density plantation in Australia. (a, b: Courtesy of Beccaro, G.; c: Courtesy of Zou, F.; d: Courtesy of Griffiths, S.)

Among the Spanish rootstocks, 'CHR-151' ('HS'), which is easily propagated via *in vitro* culture, has been widely used. For Chinese chestnut, seedling rootstock of the same scion cultivar is commonly used to avoid graft union incompatibility.

6.1.4 CULTIVARS

Important considerations for cultivar selection include medium or low vigor for medium or high-density plantings, high and consistent nut yields, pollination requirements, and susceptibility to diseases. Since tree performance and quality traits are strongly influenced by different soil and environmental conditions at each site, it is difficult to make specific cultivar recommendations (see Chapters 4 and 7 for further details).

TABLE 6.2
Rootstocks for C. sativa and Their Characteristics

Rootstock	Propagation	Compatible Cultivar	Resistant to Ink Disease	Resistant to Early Frost	Compatibility	Vigor
CA 07 Marsol	Cutting, *in vitro*	*Belle Epine, Bouche de Bétizac, Bouche Rouge, Bournette, Maridonne, Marigoule, Marrone di Chiusa Pesio, Marrone di Susa, Précoce Migoule, Marron Comball, Sauvage Marron, Précoce des Vans*. Good compatibility with most of *C. sativa* cultivar	a	c	c	c
CA 15 Marigoule	Cutting	*Maridonne, Précoce Migoule*	c	b	a	c
CA 48 Précoce Migoule	Cutting	*Bouche de Bétizac, Bouche Rouge, Bournette, Verdale (Delsol)*	a	b	N/A	b
CA 74 Maraval	Cutting, *in vitro*	*Bournette, Marron de Goujounac Maridonne, Précoce Migoule*	b	b	a	b
CA 118 Marlhac	Cutting, *in vitro*	*Belle Epine, Bouche de Bétizac, Bouche Rouge, Maridonne, Marigoule, Précoce Migoule*	b	a	b	b
CA 90 Ferosacre	Layering	*Belle Epine, Bouche de Bétizac, Bouche Rouge, Bournette, Maridonne, Précoce Migoule*. Good compatibility with most of *C. sativa* cultivar	c	a	c	b
CA 90 Ferosacre	Cutting, *in vitro*	Good compatibility with some *C. sativa* varieties	c	b	c	a
CM 904	Cutting, *in vitro*	*Martaính, Longal, Judia, Bouche de Bétizac*	b	N/A	b	b
CS1202	Cutting, *in vitro*	*Martaính, Longal, Judia, Bouche de Bétizac*	c	N/A	b	c
CS55	Cutting, *in vitro*	*Martaính, Longal, Judia, Bouche de Bétizac*	c	N/A	b	c
CM5914	Cutting, *in vitro*	*Martaính, Longal, Judia, Bouche de Bétizac*	a	N/A	b	c
111-1	Layering, Cuttings, Micropropagation	Spanish nut cultivar[d]	c	a	c	b
1483	Layering, Cuttings, Micropropagation	Spanish nut cultivar[d]	a	a	c	a

(Continued)

TABLE 6.2 (Continued)

Rootstocks for *C. sativa* and Their Characteristics

Rootstock	Propagation	Compatible Cultivar	Resistant to Ink Disease	Resistant to Early Frost	Compatibility	Vigor
2671	Cutting Micropropagation	Spanish nut cultivar[d]	a	b	c	b
7521	Layering Cuttings Micropropagation	Spanish nut cultivar[d]	a	b	c	b
1482	Layering Cuttings Micropropagation	Spanish nut cultivar[d]	c	a	c	a
7810	Layering Cuttings Micropropagation	Spanish nut cultivar[d]	b	a	c	b
Maceda_P011	Cutting Micropropagation	Spanish nut cultivar[d]	c	b	c	b
Maceda_C004	Cutting Micropropagation	Spanish nut cultivar[d]	b	a	c	a
Maceda_P043	Cutting Micropropagation	Spanish nut cultivar[d]	b	a	c	a
Maceda_P042	Cutting Micropropagation	Spanish nut cultivar[d]	b	a	c	a
Menzies	Menzies seeds	*Marrone di Chiusa Pesio, Marrone de Coppi, Red Spanish*	b	N/A	c	c

Sources: Bounous, G. *Il castagno: risorsa multifunzionale in Italia e nel mondo.* Edagricole, Bologna, Italy, p. 420, 2014; Fernández- López J, et al., O material vexetal na plantacion de Soutos, in *Guía do cultivo do castiñeiro para a producción de castañ.* Xunta de Galicia. Consellería de Medio Rural e do March, pp. 11–42, 2014[14]; Cuenca B., et al., Nuevos materiales forestales de reproducción de Castanea sp. de categoría cualificado. 6° Congreso Forestal Español. 10–14 Junio 2013. Vitoria-Gazteiz, 2013[15]; R. Costa, pers. comm.; J. Casey, pers. comm.

a = low, b = medium, c = high resistance, compatibility, or vigour, d = They are regularly grafted with many Spanish traditional varieties and 'Bouche de Betizac' but there is no information about compatibility with varieties from other countries.

FIGURE 6.2   'Précoce Migoule' flowers. (Courtesy of Gamba G.)

6.1.5   POLLINIZERS

Chestnut is a monoecious species (producing separate female and male flowers on the same tree), and the flowers develop on the current year's growth (Figure 6.2). Two types of inflorescence are found: the male catkins, occur toward the base of the shoot; and bisexual catkins towards the tip of the shoot. Female flowers appear singly or in clusters of two to four (four is common for Chinese chestnut) near the bisexual catkins. Most of chestnut varieties are self-sterile; this is not only due to the flowers maturing at different times but also to the fact that pollen and female flowers have gene incompatibility.

To achieve high nut production, pollinizers must be planted in the orchard or within 60 m of the planting. Also, for successful pollination, the cultivars selected must have synchronous bloom periods. Cultivars with long stamens (i.e., longistaminate) produce abundant quantities of pollen, while marrone type cultivars have sterile male flowers. To ensure good cross-pollination, it is important to plant more than two intercompatible cultivars. An adequate number (≥25%) of pollinizers should be planted in a uniform way in the orchard, to ensure good fruit set (Table 6.3). The chestnut is wind pollinated although insects, such as bees, may play a role in pollination.

6.1.6   PLANTING

A deep plowing before planting is usually better than digging and planting trees in holes. Plowing at a depth of 40–50 cm is recommended, but 20–30 cm is enough in shallow mountain soils. In areas with mild, moist winters, the best time to plant trees is late fall. In regions with cold winters, early spring planting is usually better. Trees are planted in berms (25–30 cm-high ridges), in holes larger than the root system (Figure 6.3). A tractor-mounted auger can be used to efficiently dig holes for tree

TABLE 6.3

Example of European Chestnut Cultivars/Pollinizers Combinations

Cultivar	Catkins	Pollinators
Belle Epine	Longistaminate	*Marigoule, Marsol, Portaloune*
Bouche Rouge	Astaminate	*Belle Epine, Marigoule*
Dorée de Lyon	Brachistaminate	*Belle Epine, Marigoule, Montagne*
Laguepie	Mesostaminate	*Belle Epine, Marigoule, Montagne*
Marron de Chevancheaux	Longistaminate	*Belle Epine, Marigoule,*
Marron de Goujounac	Longistaminate	*Marigoule, Montagne, Precoce Carmeille*
Marrone di Chiusa Pesio, di Viterbo, Fiorentino, Marradi	Astaminate	*Belle Epine, Bournette, Madonna, Marsol, Precoce Migoule, coppices*
Montagne	Longistaminate	*Precoce Carmeille*
Portaloune	Longistaminate	*Belle Epine, Marigoule, Marsol*
Precoce Carmeille	Longistaminate	*Marigoule, Montagne*
Verdale	Longistaminate	*Belle Epine; Marron de Goujounac; Montagne*

Source: Bellini, E. et al., Miglioramento genetico dell'olivo: prime osservazioni su selezioni ottenute da incrocio. *Italus Hortus*, 2, 3–12, 1995; Bounous, G. *Il castagno: Risorsa multifunzionale in Italia e nel mondo*. Edagricole, Bologna, Italy, 2014.[16]

FIGURE 6.3 Five-year-old 'Colossal' trees in a California orchard, planted in berms to repel standing water that can be conducive to root rot development. These trees were planted too close (7 × 3 m) and every other tree needed to be eliminated in later years. (Courtesy of Wilson, L.)

planting, but it should not be used when soil moisture is high, to prevent compaction at the edge of the hole. Bare root plants are often treated with a fungicide solution to prevent root diseases.

6.1.7   MULCHING

Mulches of composted sawdust, bark chips, or mowed grass, in a 1 m strip between tree rows, can be used to retain moisture, increase organic matter contents, and control weeds. For chestnut, mulching is recommended in the first years after planting. This technique reduces water evaporation, preserves the soil structure, and helps maintain an even soil temperature, favoring microflora and increasing nutrient availability. Under mulch, a shallow root system is developed exploring the more fertile parts of the soil.[5] In the USA, as in many other countries, mulch is not recommended as it provides a habitat for voles that can girdle trees by feeding on trunks during winter.

6.1.8   COVER CROPS

Cover crops may be native or sown in orchards with non-competitive legumes such as clover, or grasses such as the turf-type *Lolium perenne*, *Agrostis* spp., *Poa pratensis*, *Festuca ovina*, *F. rubra*, *F. longifolia*, and *F. arundinacea* (Figure 6.4a and b). If the cover crop is mowed and the cut vegetation left in place, the organic matter will enhance soil nitrogen content. On slopes, cover crops reduce erosion. The presence of a cover crop makes mechanical transit easier, especially after long rainy periods. Cover crops also have a positive effect on fertility, as they improve the distribution and availability of less mobile elements such as phosphorus and potassium.

6.1.9   TREE SHELTERS AND TRUNK PROTECTION

Net shelters are recommended (Figure 6.5a). The use of plastic shelters (Figure 6.5b and c) can cause burns to the bark caused by the reflection and concentration of light where solar irradiation is strong. Particularly susceptible to this bark damage is the cultivar 'Bouche de Betizac.' White plastic tree guards that spiral around tree trunks are not recommended for use year-round, as they can harbor insect pests. In some areas, the lower trunk of chestnut trees can suffer from sunscald injury. This injury occurs when the bark temperature is higher than the ambient air due to the absorption of sunlight. As the air temperature drops rapidly during the evening hours, trunk injury occurs. Where sunscald injury is common, white latex paint (diluted 50% with water) (Figure 6.6a) may be applied to lower trunks of chestnut trees to reflect sunlight and reduce the bark temperature. Painting with white latex paint or copper paint (Figure 6.6b) is also one of the ways trees are protected from rodent damage.

FIGURE 6.4 (a) Cover cropped chestnut orchard in Australia; (b) trees planted on a berm and interplanted with a cover. (a: Courtesy of Griffiths, S.; b: Courtesy of Beccaro, G.)

6.1.10 DEER CONTROL

Deer feed on terminal shoots in the spring, resulting in reduced growth. However, they are most destructive in fall and winter when males rub their antlers against young tree trunks, causing extensive bark injury or tree mortality. Newly-planted orchards are often enclosed in fencing that may be electrified where there are deer populations. Scented deer repellents are also used to protect trees, but must be reapplied as they become ineffective.

FIGURE 6.5 (a) Net shelters, (b) plastic shelters in intensive orchards in Italy and (c) Australia. (a, b: Courtesy of Beccaro, G.; c: Courtesy of Griffiths, S.)

FIGURE 6.6 (a) Sixteen-year-old 'Colossal' chestnut trees in a high-yielding Michigan chestnut orchard; the white paint reflects the sun which can cause scald injury when the sun reflects off snow during winter months; (b) copper paint applied to trunks in Italy. (a: Courtesy of Fulbright D.W.; b: Courtesy of Beccaro, G.)

6.2 WEED CONTROL

Weed control is essential during tree establishment (3–4 years). To reduce the incidence of weeds, a weed-free zone of 1–2 m under the trees is maintained. In the first year of planting, weeds can be eliminated with herbicides, surface mulching, hand weeding, or-tilling.

Organic mulches. Pine bark, saw dust, grass, straw, and wood chips are effective in conserving moisture for the plants and providing organic matter after composting.

Registered herbicides. It is hazardous to use chemical weed control in the first planting because of possible injury to the young trees unless tree trunks are shielded from spray droplets. Herbicides registered for chestnut differ among countries. Commercial products vary from pre-emergents (such as pendimethalin, oryzalin, and simazine) to post-emergents (such as glyphosate, paraquat, and diquat).

Tillage. Mechanical tillage at a 5–10 cm depth, can eliminate weeds, maintain water reserves, reduce water loss by evaporation, distribute fertilizers, aerate the soil, and promote the mineralization of organic matter. However, tillage underneath trees also causes root pruning, resulting in limited water and nutrient uptake, as well as enhanced soil erosion.

Grazing or mowing. Grazing by sheep or mowing the orchard floor is used in some production regions.

6.3   PRUNING AND TRAINING

Chestnut trees are trained to establish a strong vegetative structure before bearing nuts. New orchards are typically trained to open-center or central-leader forms. Pruning is used to increase nut size and yield, to avoid biennial bearing, and to increase light interception in the canopy. Chestnut tree inflorescences originate from the buds on 1-year-old branches. Flowers bloom from apical and sub-apical buds. Pruning of bearing trees involves the removal of old lower branches and the elimination of crowded branches to enhance light penetration within the tree canopy, resulting in large to increase nut size and quality. However, pruning reduces potential leaf surface which results in reduced root growth. For this reason, pruning should be always balanced with fertilization and irrigation.

6.3.1   Open-Center Training System

For this system, trees are trained in the first 3 years; it is used for European, Chinese, and Japanese varieties. Three to four equally spaced strong branches are selected around the trunk, which have about 120° angles, to form an open vase. At the end of the fourth year, pruning is limited to thinning cuts to enhance light interception; to eliminate dead, broken, or damaged branches; and to stimulate growth (Technical Sheets 6.1 and 6.2).

TECHNICAL SHEET 6.1   OPEN CENTER TRAINING SYSTEM

Three to four equally spaced branches are selected placed around the trunk, which have wide branch angles.

Courtesy of Facello, V.

TECHNICAL SHEET 6.2   HIGH OPEN
CENTER TRAINING SYSTEM

Many branches are selected around the trunk of Eurojapanese trees at a 1.50–2 m height.

Courtesy of Facello, V.

6.3.2   CENTRAL LEADER TRAINING SYSTEM

For high-density *C. sativa* orchards (150 or more trees/ha), central leader training is a good solution. As for many other fruit tree species, the chestnut central leader grows strongly and as near vertically as possible. The lateral branches are widely spaced around the trunk at about 45°–60° angles, to form a pyramid-shaped frame-work (Technical Sheet 6.3). The Chilean central leader, which is a variation of this training system, is described in Chapter 8.

TECHNICAL SHEET 6.3   CENTRAL LEADER TRAINING SYSTEM

Lateral branches are widely spaced around the trunk.

Courtesy of Facello, V.

6.4   FERTILIZATION

6.4.1   BEFORE PLANTING

Nutrient recommendations are based on local site conditions. Based on soil test results, manure and/or mineral fertilizers are often incorporated into the soil before tree planting. For chestnut orchards, the recommended soil organic matter content is at least 2%, When organic matter is low, manure at generally 40–50 t/ha can be incorporated into the soil at a 20–30 cm depth by plowing. Mineral fertilizers with phosphorus and potassium can also be soil-incorporated before planting. Generally, superphosphate and potassium sulfate fertilizers are used, but potassium chloride may be an alternative. A content of up to 150 mg kg⁻¹ soil, potassium promotes water and heat stress resistance and the growth of nuts. When boron is low (less than 1 mg kg⁻¹ soil) in this element, the application of this nutrient may enhance nut production.

Generally, for non-bearing trees in Italy, 50 g/plant of nitrogen in the first year with applications, increasing to 200–250 g/tree by the third to the fifth years, along with potassium at 80 g/tree. Nitrogen is often applied to the soil surface in the form of ammonium sulfate or ammonium nitrate to stimulate vegetative growth in the spring. From the sixth year, depending on the number of fruits harvested and soil conditions, fertilizer should be applied at the following rates: N (60–80 kg/ha); P (9–13 kg/ha); K (66–100 kg/ha). If a complex fertilizer is used, it should have a low P content and no chlorides. For example, 150–200 g/tree of 15–9–15, 15–5–20, or 12–6–27 fertilizer or a similar product can be applied to a soil with an average mineral content.

6.4.2   BEARING ORCHARDS

Nitrogen. The nitrogen needed depends on soil type, chestnut species and cultivar, and the age and productivity of the trees. For example, for *C. sativa* and Eurojapanese hybrids, from the sixth year onward, approximately 60–80 kg/ha/year of nitrogen should be applied (e.g. 0.3–0.4 t/ha of ammonium nitrate), which will supply the tree needs and that of a cover crop.

Phosphorus. Depending on the site and the intensity of production, it is not usually necessary to add phosphorus before the tenth year.

Potassium. Deficiencies in potassium are more common in light-textured soils and, when it is detected, recommended applications range from 25 kg/ ha for a 6-year-old tree to 120 kg/ha for 10-year-old trees.

When using complex fertilizers (e.g. 15–9–15, 15–5–20, or 12–6–27 or similar), 800–900 g/tree/year are usually applied.

6.4.3   WHEN TO APPLY NITROGEN

During the adult phase, nitrogen is one of the most limiting nutrients and is applied twice: once in early spring, and a second time in late spring. In North America, nitrogen fertilizer is applied twice during the growing season. Half of the required amount is applied at bud swell (about April 1) in the spring, and the other half at the end of flowering (about June 20).

6.4.4   EMPIRIC UPTAKE-BASED ACCOUNTING

As for many other fruit crops, the nutrients needed can also be calculated based on the amounts of mineral elements utilized annually by parts of the chestnut tree. The mineral elements needed for re-integration can be estimated as follows:[6]

$$M_{UT} = M_{SB} + M_{LE} + M_{CH} + M_{FL} + M_{BU} + M_{SR} + M_{AST}$$

M_{UT}, Mineral element uptake
M_{SB}, Mineral element in small branches
M_{LE}, Mineral element in leaves
M_{CH}, Mineral element in chestnuts

M_{FL}, Mineral element in flowers
M_{BU}, Mineral element in burs
M_{SR}, Mineral element in small roots
$M_{\Delta ST}$, Mineral element in structural roots and branches increase

Each tree part that does not return to the orchard (e.g. fruits, leaves, pruned branches) can be converted into nutrient uptake from the soil and used to determine how much fertilizer to apply. Technical Sheet 6.4 presents the mineral element concentration in leaves, branches, and flowers that can be used to estimate the macro- and micronutrient uptake by the chestnut orchard. In adult trees the trunk biomass increase is very limited, thus it can be assumed that biomass is stable over time. Moreover, nutrients absorbed and distributed in roots are hard to quantify, but this flow can be partially compensated by root mortality and rhizodeposition.

TECHNICAL SHEET 6.4   HOW TO CALCULATE FERTILIZER DOSE BASED ON CHESTNUT UPTAKE

1. Where are mineral elements stored in a chestnut tree?
   Which stay or return to the orchard?
   Which plant part is removed from the orchard?

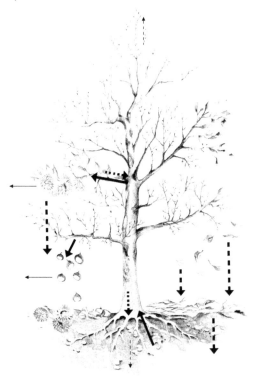

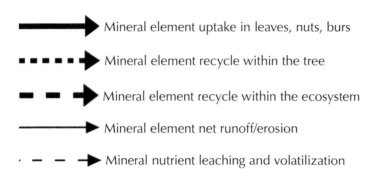

Mineral element uptake in leaves, nuts, burs

Mineral element recycle within the tree

Mineral element recycle within the ecosystem

Mineral element net runoff/erosion

Mineral nutrient leaching and volatilization

Courtesy of Facello, V.

2. Mineral elements stored in the chestnut tree organs (kg/t DW)

	C	N	P	Ca	Mg	K	Na	Mn	Fe	Cu	Zn
	(kg · t⁻¹ DW)						(g · t⁻¹ DW)				
Leaves	470–480	12–14	0.8–1.2	6–7	2–3	2.5–3	245–255	650–680	95–100	12–17	20–25
Branches	470–480	8–9	0.5–0.8	7–8	0.8–1.7	1.8–2.2	260–270	350–370	100–105	6–8	30–35
Burs	480–490	13–15	0.5–1.5	5–7	0.8–1.6	2–4	230–240	360–380	160–170	10–14	19–23
Flowers	470–490	16–18	0.8–1.2	3–4	1.5–2.5	5–7	210–220	330–350	105–115	12–16	20–23
Fruits	450–470	8–12	0.8–1.3	3–4	0.8–1.2	7–8	230–240	165–185	30–50	10–12	15–17

Source: Average *C. sativa* values. Data DISAFA, University of Turin, Italy.

3. How many macro- and micronutrients do you take out in a chestnut orchard depending on fruit production?

Orchard producing 2 t · ha⁻¹ of chestnuts (fresh weight) per year

	C	N	P	Ca	Mg	K	Na	Mn	Fe	Cu	Zn
	(Kg · ha)						(g · ha⁻¹)				
Burs (if removed)	245–300	5–10	0.5–1	2.5–3	0.5–1	1.5–2	120–145	195–235	85–105	6–8	11–15
Chestnuts	440–530	10–15	1–1.5	3–4	1–1.5	7–9	225–270	165–200	35–50	11–15	15–20
TOTAL	**685–830**	**15–25**	**1.5–2.5**	**5.5–7**	**1.5–2.5**	**8.5–11**	**345–415**	**360–435**	**120–155**	**17–23**	**26–35**

Source: Average *C. sativa* values. Data DISAFA, University of Turin, Italy.

6.4.5 SILICON APPLICATION

The impact of abiotic stresses such as drought and heat have intensified in recent years due to climatic changes, interfering with plant robustness and so with plant health and productivity. In trials where European chestnut seedlings were fertilized with silicon, an increase in their resilience to abiotic stress was observed. Under drought conditions, silicon helped plants to maintain a more adequate water potential

TABLE 6.4

Chestnut Leaf Analysis Reference Ranges (*C. sativa*, % Dry Matter)

Nutrient	N	P	K	Ca	Mg
Reference range	2.2–2.9	0.1–0.3	0.5–2.2	0.5–1.4	0.2–0.7

and, consequently, the photosynthesis rate. Silicon can act as an anti-stress agent and prevent structural and functional deterioration of cell membranes induced by abiotic stresses. Additionally, in fertilized plants, the capacity for recovery was also increased. These results suggest silicon fertilization may protect chestnut trees from environmental stress, especially as in actual climate change occurs.[7]

6.4.6 UNDERSTANDING DEFICIENCIES WITH LEAF ANALYSIS

Foliar analysis is used to estimate the nutritional status of chestnut trees. Sufficiency ranges for each nutrient are used to provide fertilizer recommendations. The sufficiency range indicates the values at which the tissue is at the optimal nutritional status. Sufficiency ranges vary by *Castanea* species, and fertilizer recommendations based on test results are often provided locally. Nutrient reference ranges in chestnut leaves are reported in Table 6.4. For foliar analysis, a sample of 50 fully expanded leaves from the mid-shoot portion of non-fruiting shoots is collected from July 15 to August 15.

6.5 IRRIGATION MANAGEMENT

6.5.1 WHY, HOW MUCH, AND WHEN TO IRRIGATE

Why irrigate chestnut? During orchard establishment, irrigation is critical due to the tree's limited capacity to absorb soil moisture from fine root hairs and small feeder roots. Under agroforestry conditions, the chestnut tree is mainly dependent on rainfall and on its capacity to develop an extensive root system. Irrigation is one agricultural practice that promotes high yields. Because water resources are often limited, water management is based on the soil-plant-water relationship. The resilience of chestnut to drought is weak if the tree is not planted in a deep soil with a high water retention capacity, or if it is infected with root diseases. Irrigation will likely become increasingly important with the tendency of summers to be even longer and drier.

How much water? Chestnut trees require about 800–1300 mm/year depending on the species.[2,9] Eurojapanese hybrids need 1200–1300 mm/year. However, even more important than the total annual precipitation is the distribution of rainfall throughout the growing season. For a 1- to 2-year-old Eurojapanese hybrids orchard in dry conditions, from 15 to 30 L of water tree per/week is needed to ensure good vegetative growth. Few irrigation studies have been conducted, and results vary due to tree species, rootstock, cultivar, tree age, and soil-climatic conditions. In trials conducted by Mota et al.,[10] the final chestnut weight was 34% higher in irrigated orchards compared with non-irrigated orchards. In any case, irrigation adds commercial value to chestnuts.

When does chestnut need water? Bearing chestnut trees need water during two phenological stages: flowering and fruit growth. When exponential fruit growth occurs (at the end of the summer/beginning of autumn), climatic conditions can be dry and hot, and chestnut production can be limited. From the compilation of many studies, chestnut production is higher in irrigated orchards that receive 180 mm of moisture between June and September.

6.5.2 SMART IRRIGATION TECHNIQUES IN CHESTNUT

Which irrigation system? For proper water management, the equipment, amount, and timing of irrigation are important. In chestnut orchards, the irrigation systems most commonly used are the drip (Figure 6.7) and the micro-sprinkler types. The type of system used depends on water availability and purity, soil permeability and water storage capacity, topography, energy costs, capital, and technology requirements.[11] The micro-sprinkler system promotes a larger wetted area, but it can also promote the dispersion of *Cryphonectria parasitica* if it also wets the trees trunk. Thus, drip irrigation may be preferable in areas where this disease is prevalent.

Using devices in the chestnut orchard. In modern fruit orchards, irrigation scheduling is based on values from monitoring devices and sensors that trigger a remotely controlled irrigation system. In chestnut trees, the maximum photosynthetic rate was registered during midday until stem water potential reached −1.2 MPa. The reference soil water content value to trigger irrigation was below 16% volume in the study by Mota et al.[12] If irrigation scheduling is based on sensors that reflect the soil matric water potential, then irrigation should start every time that it is below −100 cbar.

With these references, the chestnut producer can manage irrigation properly. However, soil reference values vary, depending on the type of soil and their soil retention curves. For example, a fine sandy-loam soil at −100 cbar has about 17% water volume, whereas the same water tension value for a sandy soil has only 5% volume water content. Additionally, soil properties such as field capacity (the amount of water that a well-drained soil should hold against gravitational forces) and wilting point (soil water content at which plants will permanently wilt because the crop can no longer extract the remaining water) should be known for proper soil water management.

FIGURE 6.7   A drip irrigation system in a 10-year-old orchard. (Courtesy of Beccaro, G.)

6.6 HARVEST

Ripening. Chestnut trees typically bear their first crop 3–4 years after planting. Chestnuts are collected when they drop from the tree. In some cultivars, dropped nuts remain enclosed in burs; in others the chestnuts drop to the soil from open burs still hanging on the branches. Ripening is gradual and can occur over one month. In warmer zones, it begins at the end of summer for early bearing cultivars, to autumn (November in the NE) for late cultivars. In dry years, burs opening is either delayed or it never occurs. Chestnuts harvested before complete ripening are difficult to store. The harvest period is relatively short, with nuts collected every few days. High temperature, moisture, pathogens, and predators can result in poor nut quality.

Harvest methods. Chestnuts may be hand harvested or mechanically harvested. When hand harvested, nuts dropped on the ground may be separated from burs using hammers or gloves. In some countries, branches are struck with wooden poles to help nuts fall, but this practice should be avoided because it wounds branches. Hand harvesting is generally expensive (depending on country and labor cost, up to 50% of the total production cost) because pickers harvest a small number of nuts/hr (about 10–15 kg/hr).

Roll-out nets are used in France and Corsica to efficiently collect dropped nuts and burs (Figure 6.8). Nets can be laid on the soil or supported by poles at 1.2–1.6 m above the soil. Rolling out the nets is time-consuming, but it permits pickers to operate with greater efficiency. The cost of the nets is high, depending on total or partial covering of the orchard.

Mechanical harvesting in the chestnut industry is becoming more common. A number of problems have been solved recently, such as abrasions on the shell of the nuts. Harvesting machines, vacuums, and sweepers (Figures 6.9, and 6.10), similar to those used for walnuts, hazelnuts, and almonds have been adapted for chestnut and are able to separate nuts from burs. In North America, a hand-held tool (Nut-Wizard; Figure 6.11) is often used to pick up the nuts in most orchards <4 ha. Alternatively, a cost-effective modified paddock vacuum can be used in small-scale orchards for nut collection and sorting burs (Figure 6.12). Pecan harvesters can be used for chestnuts, but these machines fail to pick up some of the nuts embedded in the cover crop or soil due to the flat side of nuts.

FIGURE 6.8   Roll-out nets used in Ardèche (France) to collect the nuts. (Courtesy of Beccaro, G.)

FIGURE 6.9   Vacuum machine for mechanical harvesting. (Courtesy of Beccaro, G.)

FIGURE 6.10   Sweepers machine for mechanical harvesting. (Courtesy of Monchiero & C. s.n.c.)

FIGURE 6.11   Chestnut harvest using a Nut Wizard in USA. (Courtesy of Warmund, M.)

FIGURE 6.12   Modified paddock with sorting table. (Courtesy of Warmund, M.)

REFERENCES

1. Mellano, M.G., Cerutti, A.K. and Beccaro, G.L. 2016. High density chestnut cultivation: Aspects of sustainability. *Castanea* 5: 8–9.
2. Bounous, G. and Beccaro, G.L. 2002. Chestnut culture: Directions for establishing new orchards. *FAO – CIHEAM, Nucis* 11: 30–34.
3. Pereira-Lorenzo, S. and Fernández-López, J. 1997. Description of 80 cultivars and 36 clonal selections of chestnut (*Castanea sativa* Mill.) from Northwestern Spain. *Fruit Varieties Journal* (USA), 51: 13–27.

4. Fernández-López, J., Vazquez-Ruiz-de-Ocenda, R.A., Díaz-Vázquez, R. and Pereira-Lorenzo, S. 2001. Evaluation of resistance of *Castanea* sp. clones to *Phytophthora* sp. using excised chestnut shoots. *Forest Snow and Landscape Research* 76: 3–451–454.

5. Måge, F. 1982. Black plastic mulching, compared to other orchard soil management methods. *Scientia Horticulturae* 16(2): 131–136.

6. Grignani, C. 2016. *Fertilizzazione Sostenibile*. Edagricole, Bologna, Italy, p. 444.

7. Zhang, C., Moutinho-Pereira, J.M., Correia, C., Coutinho, J., Gonçalves, A., Guedes, A. and Gomes-Laranjo, J. 2013. Foliar application of Sili-K® increases chestnut (*Castanea* spp.) growth and photosynthesis, simultaneously increasing susceptibility to water deficit. *Plant and Soil* 365(1–2): 211–225.

8. Warmund, M.R. 2018. Nutrient status and fruiting response of young chinese chestnut trees following application of nitrogen. *Journal of the American Pomological Society* 72(1): 12–20.

9. Beccaro, G.L., Mellano, M.G., Barrel, A. and Trasino, C. 2009. Restoration of old and abandoned chestnut plantations in Northern Italy. *Acta Horticulturae* 815: 185–190.

10. Mota, M., Pinto, T., Marques, T., Borges, A., Caço, J., Raimundo, F. and Gomes-Laranjo, J. 2018a. Study on yield values of two irrigation systems in adult chestnut trees and comparison with non-irrigated chestnut orchard. *Revista de Ciências Agrárias* 41(1): 236–248.

11. Pereira, L.S. 2004. *Necessidades de água e métodos de rega*. Publicações Europa-América, 1st ed, Lisboa, Portugal.

12. Mota, M., Marques, T., Pinto, T., Raimundo, F., Borges, A., Caço, J. and Gomes-Laranjo, J. 2018b. Relating plant and soil water content to encourage smart watering in chestnut trees. *Agricultural Water Management* 203: 30–36.

13. Bounous, G. 2014. *Il castagno: Risorsa multifunzionale in Italia e nel mondo*. Edagricole, Bologna, Italy, p. 420.

14. Fernández- López, J., Miguel-Soto, B., Miranda-Fontaiña, M.E., Fernández-Cruz J., Barciela-García S. and Martínez-Picos R. 2014. Omaterial vexetal na plantacion de Soutos. In: *Guía do cultivo do castiñeiro para a producción de castañ*. *Xunta de Galicia*. Conselleria de Medio Rural e do Mar, La Coruna, pp. 11–42.

15. Cuenca, B., Luquero, L. and Ocaña, L. 2013. *Nuevos materiales forestales de reproducción de Castanea sp. de categoría cualificado*. 6º Congreso Forestal Español. 10–14 Junio 2013. Vitoria-Gazteiz

16. Bellini, E., Parlati, M.V., Pandolfi, S., Giordani, E., Perri, E. and Silletti, A. 1995. Miglioramento genetico dell'olivo: Prime osservazioni su selezioni ottenute da incrocio. *Italus Hortus* 2(1–2): 3–12.

7 Chestnut Farming with Chinese, Japanese, and Eurojapanese Hybrid Cultivars

Feng Zou, Sogo Nishio, Michele Warmund,
Gabriele Beccaro, Giancarlo Bounous,
and Michele Bounous

CONTENTS

7.1 CHINESE CHESTNUT

Many high-density plantations are established with *Castanea mollissima* outside China, due to its relatively small tree size and high productivity.[1] The species is relatively blight-resistant although it can be affected by *Cryphonectria parasitica*. *C. mollissima* is adaptable to a wide range of climatic and soil conditions. Chinese chestnut trees can tolerate temperatures as low as −30°C in mid-winter when fully dormant. However, when temperatures drop rapidly in the fall before trees are fully cold-acclimated, terminal shoot dieback can occur. Chinese chestnut floral buds are also susceptible to spring frost injury, especially when trees are planted at low-lying sites where cold air can drain from the surrounding area (i.e. a frost pocket). Chinese chestnut requires about 1000–2000 mm/year of rainfall and a mean annual temperature of 10.5°C–21.8°C, and in China it can be planted at up to 2800 m.[2] In recent years, climate warming has affected chestnut trees phenology throughout northern China (Figure 7.1). Photosynthetically active radiation (PAR) has decreased significantly in Beijing over the past 50 years (Figure 7.2). Reduced PAR between September 24th and February 5th has had an impact on Chinese chestnut flowering, and it can partially explain a modified trend in the flowering period.[3]

For *C. mollissima*, harvest begins in early September in northern China, whereas most cultivars bear fruits in October in southern China.[4] Trees generally drop nuts for 2 to 4 weeks. In China, most growers mainly gather chestnut burs with bamboo poles. In larger orchards, mechanical harvesters are replacing this practice.

FIGURE 7.1   Modern *C. mollissima* plantation in northern China. (Courtesy of Zou, F.)

FIGURE 7.2   Old *C. mollissima* tree in Beijing City. (Courtesy of Zou, F.)

7.1.1   ORCHARD LAYOUT

Chinese chestnut trees are typically planted at 2.0×4.0 m to 3.0×4.0 m spacing. Chinese Chinquapin trees (*C. henryi*) are planted at 3.0×4.0 m to 4.0×4.0 m in the hilly regions of southern China. The planting density ranges from about 630 to 1600 trees/ha. Several cultivars that bloom at the same time are planted together. For example, *C. mollissima* cultivars 'Qianxi15' and 'Dabanhong' are the best pollinators for 'Yanshanzaofeng'; 'Qianxi14', 'Zunyu', and 'Qianxi15' are the best pollinators for 'Dabanhong' in the Yanshan Mountain region of Northern China.[5] *C. henryi* cultivar 'Huali No. 3' is a good pollinator for 'Huali No. 1', 'Huali No. 2', and 'Huangzhen'.[6] The ratio of main cultivar/pollinizer cultivar is usually 4 ~ 5: 1 (Figures 7.3 and 7.4).

In North America, Chinese chestnut trees are planted at lower densities (69 to 123 trees/ha), especially at sites with fertile soils. Typical plant spacings are 9×9 m up to 12×12 m. Often, one row of a pollinizer cultivar is planted between three rows of another cultivar for ease of harvest. Individual trees are staked (1.8 m-tall) to support the central leader. Stakes are maintained in the orchard for about 5 years. By this time, trees are trained to a central leader system with a strong framework. A sod cover is used between tree rows, using turf-type fescues (*Festuca arundinacea, F. rubra, F. longifolia*) or a mixture of 75% turf-type fescue and 25% Kentucky bluegrass (*Poa pratensis*). A vegetation-free strip (1.8 m-wide) is maintained under the trees (Figure 7.5).

7.1.2   CULTIVARS AND ROOTSTOCKS

7.1.2.1   Chinese Cultivars

More than 300 *C. mollissima* cultivars and ecotypes are described in China and, usually, the sub-tropical types bear larger nuts than northern ones. Larger nuts are processed and exported. The fruit size varies from 10 to 30 g/nut, and the nuts are usually easy

FIGURE 7.3   Heavy bearing *C. mollissima* orchard. (Courtesy of Zou, F.)

FIGURE 7.4   *C. mollissima* and herbaceous crops consociation. (Courtesy of Zou, F.)

to peel. Cultivars ripening in September have a niche market and receive the best price in northern China. Popular cultivars in northern China are 'Yanshanzaofeng', 'Yankui', 'Dabanhong', 'Yanchang', 'Yanjin', 'Yanshanhong', 'Shifeng', 'Daiyuezaofeng', 'Hongli', 'Zhenandabanli', 'Chenguoyouli', 'Xinyangdabanli', 'Queshanziyouli', 'Dahongpao', etc. Good cultivars in southern China are 'Jiujiazhong', 'Chushuhong', 'Jiaozha', 'Qingzha', 'Chili', 'Hongmaozao', 'Kuili', 'Maobanhong', 'Baopidayouli', 'Tanqiaobanli', 'Huaqiaobanli', 'Shaoyangtali', 'Jiebanli', 'Shaoguan 18', 'Zhongguohongpili', 'Yunyao', 'Yunzao', etc.[2]

FIGURE 7.5 Four-year-old 'Qing' trees on AU-Cropper rootstock in Missouri. (Courtesy of Fulbright, D.W.)

FIGURE 7.6 *C. henryi* orchard in southern China. (Courtesy of Zou, F.)

More than 20 *C. henryi* cultivars and ecotypes are grown in Southern China.[7,8] Fruit size varies from 6 to 15 g/nut. The nut starch content varies from 25.7% to 55.1%. Popular cultivars are 'Huali No. 1','Huali No. 2','Huali No. 3','Huali No. 4', 'Bayuexiang', 'Zaoxiangli', 'Jianou No. 1', 'Jianou No. 2', 'Bailuzi', 'Huangzhen' and 'Youzhen' (Figure 7.6).

C. seguinii is used on its own roots as a food and timber crop. Nuts are small and brown, and the weight of a single nut is about 1 g. The nut has low commercial value and is usually used as rootstocks for *C. henryi* and *C. mollissima*.[2] Nuts are eaten as a traditional food in China, where they are consumed fresh, cooked, candied, and as a source of flour for pastries.

Popular rootstocks in China are the seedlings of *C. mollissima* and *C. seguinii*.

7.1.2.2   U.S. Cultivars

In North America, about 75 Chinese chestnut cultivars are available. However, 'Gideon', 'Qing', 'AU-Homestead', and 'Sleeping Giant' are recommended for planting in the mid-western region of the USA. Of these cultivars, three have a spreading growth habit, whereas 'Sleeping Giant' has an upright habit. Trees grown in Missouri generally produce the first flowers (known as primary flowers) in May and June. Another period of flowering occurs in late July and August (secondary flowers) on several Chinese chestnut cultivars, but this bloom occurs too late in the growing season to produce marketable nuts of sufficient size before a hard frost occurs (−5°C) (Figure 7.7).

Grafted trees generally begin bearing nuts 3 years after planting. Harvest occurs from mid-September through to mid-October. Typically nut yields in well-managed low-density orchards are about 2,200 kg/ha when the trees are 12 to 15 years old.

FIGURE 7.7   Primary and secondary burs and flowers on a 'Williamette' chestnut shoot. (Courtesy of Warmund, M.)

Trees propagated by seed are also planted by some producers to reduce tree costs, but these orchards are less precocious and often produce lower nut yields than those with grafted trees.

Cultivars are grafted onto seedling rootstock. Because delayed graft incompatibility is common for Chinese chestnut cultivars used in North America, a seedling rootstock of the same scion cultivar is often used. Alternatively, 'AU-Cropper', a vigorous rootstock, has been used with several scion cultivars, as well as less vigorous 'Qing' rootstock. However, graft union incompatibility resulting in tree mortality can occur 4 to 7 years after grafting 'Qing' with some scion cultivars. 'Little Giant', a complex hybrid of *C. mollissima* × *C. seguinii*, is a dwarfing rootstock or interstem that is compatible with some Chinese chestnut cultivars, but is not yet used commercially.

7.1.3 ORCHARD MANAGEMENT

7.1.3.1 Training and Pruning

Chinese chestnut trees are trained and pruned to maximize the sunlight interception. New plantings in China are typically trained to a low-height open-center or to a central-leader forms. Open-center trees allow sunlight into the vase-shaped middle, as well as the tops and sides.

7.1.3.2 Chinese Open-Center

The open-center training system used in China with *C. mollissima* is quite different from that used elsewhere. Initially, a central leader is selected with two or three lower branches. In the third year, outward growth is encouraged and inward-growing branches are removed. For this training system, commercial cropping begins in approximately the fifth growing season. The central leader is then cut out in the winter, leaving two to three permanent branches. In the following years, interior branches are thinned when they become as they begin to crowded, and branches are pruned to maintain a reduced tree height and elliptical tree shape. Wang et al. (2017) found that a 30% pruning treatment for 7-year-old *C. mollissima* trees significantly improved yield.9 Excessive pruning promotes vegetative growth but diminishes and delays nut production (Figure 7.8a and b).

7.1.3.3 Chinese Chestnut Central Leader

This training system is used in the USA. Trees are trained and maintained in a pyramidal shape with a dominant central leader to maximize light penetration to the tree canopy. The ideal tree from the nursery has a central leader with at least four lateral branches spaced evenly around the trunk. These lateral branches should be located between 60 and 80 cm from the soil surface and have wide branch angles. If the tree has few lateral branches at planting, they are cut back to the trunk and the leader is pruned at about 80 cm from the soil surface at planting to promote the growth of new lateral branches. The leader is also tied to a stake at this time. When the central leader has grown to about 40 cm-long, the first, second, and third shoots located near the tip of leader are removed, leaving the lower branches to develop into scaffold limbs. During the first dormant season, four scaffold branches, equidistantly

FIGURE 7.8 (a, b) *C. mollissima* trees trained to an open-center form in southern China. (Courtesy of Zou, F.)

spaced around the trunk, are selected and branches, especially those with narrow crotch angles or bark inclusions, are removed by pruning. Until trees begin bearing, they are lightly pruned to maintain the central leader and competing branches are removed. As trees mature, overlapping branches are pruned and thinning cuts are made at the trunk to enhance light interception and promote cropping. Also, some lower limbs may be removed to facilitate harvest underneath the tree canopy.

7.1.3.4 Nutrient Management

In China, most chestnut growers apply 15–20 kg per tree of organic fertilizers such as manure before planting trees. In spring, KH_2PO_4 and boron fertilizers are usually

FIGURE 7.9   *Lolium perenne* as cover crop in a Chinese chestnut orchard. (Courtesy of Zou, F.)

applied at flowering. In summer, 0.5–1.0 kg of $(NH_2)_2CO$ and 0.5–0.8 kg of KH_2PO_4 per tree are applied as a top dressing in adult trees.[10] However, the optimum nitrogen rate varies depend on the soil type, soil management, and crop load.

In North America, about a month after planting, 70 g of nitrogen is applied underneath the tree, with the same amount applied about 6 weeks later. In subsequent years, the amount of nitrogen applied to non-bearing trees is based on test results and yield goals, with nitrogen rates ranging from 56 to 140 kg/ha when trees are mature.

C. mollissima has superficial roots that can easily become water-stressed and limit nutrient absorption in dry regions; therefore, irrigation is recommended (Figure 7.9).

7.2   JAPANESE CHESTNUT

7.2.1   GENERAL INFORMATION

Japanese chestnut is a small tree and most cultivars produce nuts at 3 years of age, when they are 2 m tall. For *C. crenata* trees, bud-break, flowering, and harvesting occur earlier than in Chinese and European chestnut. Although wild trees have small nuts, cultivars generally produce very large nuts (>30 g). Japanese chestnut trees have been cultivated in hilly and mountainous areas in the past, but production in these areas has steadily decreased. Most of the new plantations in Japan are designed as fruit tree orchards on flat land (Figure 7.10a and b). The mean annual temperature in areas of cultivation is 7°C to 17°C. Producers consider the life span of cultivated trees to be about 15 years. After this time, old trees are cut down and replaced with new ones.

(a)

(b)

FIGURE 7.10 Modern intensive Japanese chestnut orchards in Ibaraki Prefecture (a, b). (Courtesy of Nishio, S.)

7.2.2 ORCHARD LAYOUT

Japanese chestnut trees are typically planted at a spacing from 5.0 × 5.0 m to 9.0 × 9.0 m. When a producer plants a new orchard, usually twice as many trees as needed are planted in each row. When the trees mature and become crowded, the filler trees are removed (see Section 7.2.4.1). Japanese chestnut cultivars are generally self-sterile so cross-pollination is recommended. *C. crenata* pollen density declines remarkably when a producing tree is more than 10 m from the pollinizer.

7.2.3 CULTIVARS AND ROOTSTOCKS

The Japanese cultivars are generally more exigent than European and Chinese ones in terms of climate, type, and fertility of soil. The cultivars tend to be precocious and produce large nuts with a marked hilum. Also, Japanese chestnuts are generally not as sweet as European ones. The quality of nuts is good only at lower altitudes when trees are grown with good cultural practices and pruned annually. The cultivars show some resistance to canker blight and to ink disease, which is one of the main reasons for their introduction in Europe.

'Ishizuki', 'Tsukuba', 'Tanzawa', and 'Ginyose' produce good-sized nuts, but productivity in Europe is lower than in Japan. Different trials suggest careful management when grown in large-scale orchards out of their original area.

Most Japanese chestnut cultivars have a pellicle that is difficult to peel, which increases the labor and cost of removing it from the nuts during processing. Because of the difficult peeling, the surface of the nut is usually cut off along with the pellicle, resulting in a low price for small nuts. However, easy-peeling cultivars 'Porotan' and 'Porosuke' were recently released by the Institute of Fruit Tree and Tea Science, NARO (National Agriculture and Food Research Organization).[11] When easy-peeling cultivars are grown, it is important to avoid contamination by nuts from difficult-to-peel cultivars. One possible solution is interplanting of only easy-peeling cultivars. Another solution is to plant 'Riheiguri', which has a clearly different appearance, or 'Ishizuki' and 'Mikuri', which have a different harvesting time from 'Porosuke' and 'Porotan', as pollinizer trees. Although the pellicle of Japanese-Chinese hybrids is difficult to remove when they are pollinated by Japanese chestnut cultivars,[12] nut peelability of 'Porotan' is good even if it is pollinated with the major chestnut cultivars in Japan.[13]

Seeds of major *C. crenata* cultivars are generally used as rootstocks. Some cultivars show graft–rootstock incompatibility, resulting in decreased tree vigor. In such cases, using a rootstock of the same cultivar as the scion or genetically similar to it is the best way to avoid incompatibility. Clonal rootstocks have been tested in Italy for *C. Crenata*, and 'Marlhac' has good compatibility with 'Ishizuki'.

7.2.4 ORCHARD MANAGEMENT

7.2.4.1 Training and Pruning

Although Japanese chestnut is considered a crop that requires low amount of labor, it is necessary to control the number of branches and to harvest large nuts to make a good profit. *C. crenata* produces strong branches when it is less than 5–7 years old. During this time, many producers prune heavily, even if yields are expected to be low, and they usually get a better return than other producers. As for other *Castanea* species, Japanese chestnut is intolerant to shade (i.e., a mother shoot exposed to little sunshine sets few burs). Thus, annual pruning is recommended. In the past 30 years, low-tree-height tree training has been introduced to many farmers to facilitate pruning (Figure 7.11a–c). This method limits tree height to less than 3.5 m and reduces the density of mother shoots that are >6 mm in diameter (measured in the basal part)

FIGURE 7.11   (a–c) *C. crenata* trees trained to an open-center form in Ibaraki prefecture. (Courtesy of Nishio, S.)

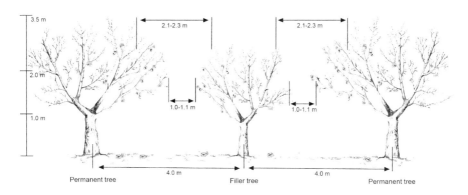

FIGURE 7.12   Hedging of temporary filler trees. (Courtesy of Beccaro, G.)

to 5–7/m². [14] After 10 years, filler trees are removed or severely cut back. Branch-to-branch distance is limited to 1–1.1 m in the lower part of the trees and 2.1–2.3 m in the upper part (Figure 7.12).

In Japan, weeding is usually carried out mechanically at least three times per year, in May, in late July, after the end of the rainy season, and in late August, just before harvest. [14]

7.2.4.2   Nutrient Management

In Japanese chestnut orchards, when trees are more than 10 years old, annual fertilization generally consists of: 120–200 kg N/ha, 80–140 kg P_2O_5/ha, and 120–200 kg K_2O/ha. [15]

The fertilization requirements for 3- to 4-year-old and 6- to 7-year-old trees are one-third and two-thirds, respectively, of those for 10-year-old trees. Fertilization is performed three times per year. About half of the total fertilizer amount is applied in January to March, another 20% in June, and the remaining 30% in September to October. Although about 10% livestock waste compost is generally provided to the soil when grafted seedlings are planted, growers in areas prone to freezing injury avoid adding it to compost because the freezing tolerance of trees decreases under excessive N conditions. [16] Both Japan and Korea are rich in rainfall, so irrigation is not necessary in those regions.

7.2.4.3   Specific *C. crenata* Physiological, Pest and Diseases Problems

Japanese chestnut trees are susceptible to freezing injury, including spring frost, and black root rot (*Macrophoma castaneicola* and *Didymosporium radicicola*). Spring frost and freezing injury are very serious problems for *C. crenata*. Both frost injury and a disease are often observed, and producers usually do not know exactly what has caused tissue dieback. To prevent withering, seedlings grafted at higher than 30 cm above the ground are planted.

Peach moth (*Conogethes punctiferalis*) oviposits on burs and nuts, and its purple larvae attack the nut at harvesting time. Peach moth has three generations per season, and the third generation generally attacks the nut of early-ripening cultivars. Some producers use insecticides two or three times, e.g., before harvest (third generation) or after flowering (first generation). The infestation rate of the nuts differs greatly depending on the area and conditions. If the infestation rate is low, producers often avoid the use of insecticides.

The larvae of chestnut weevil (*Curculio sikkimensis*) infest late-ripening cultivars. Even when nuts seem to be normal at harvesting time, larvae can emerge during transportation and storage. For this reason, methyl bromide fumigation was once used to kill the larvae in the egg just after harvesting. Methyl iodide is used as a substitute, but it is not easy to use and is more expensive, as it requires a dedicated facility. The lack of a safe, convenient, and effective chemical is a serious problem for cooperatives who want to fumigate before selling. Alternative disinfestation methods include refrigeration, hot-water treatment, and or the application of insecticides. Refrigeration at −2°C to 0°C for 4 weeks or hot-water treatment at 50°C for 30 minutes can kill most of the larvae. However, large-scale equipment for refrigeration or hot-water treatment is currently cost-prohibitive for Japanese cooperatives. Using insecticides 2 weeks before harvest effectively reduces infestations, but is not yet widely popular.

White-striped longhorn beetle and swift moth (*Batocera lineolata* and *Endoclyta excrescens*) attack tree trunks and are also a severe problem for Japanese chestnut. Crushing the oviposition scars and pouring pesticide into the holes dug by these pests are effective in reducing the damage. Regular mowing can suppress the swift moth population because this insect also infests annual species.

7.3   EUROJAPANESE HYBRIDS

7.3.1   GENERAL INFORMATION

Eurojapanese hybrids are very suitable for cultivation in orchards, and the trend for establishing new plantings is increasing every year, mainly for three reasons: early nut ripening, large nuts and easy tree care. Their nuts, which ripen earlier than European ones, are more suitable for fresh consumption than for industrial processing hybrids should not be planted above 500–700 m. Generally low- and mid-mountain hilly areas are preferred, well exposed to sun and not subjected to spring frosts. In spite of late bud-break (March-April), the tender shoots of hybrids may be damaged.

7.3.2 ORCHARD LAYOUT

Tree spacings ranges from 6 × 8 m (208 plants/ha) 8 × 10 m (125 plants/ha); the distance varies according to tree vigor, site, and cultural practices.

7.3.3 CULTIVARS AND ROOTSTOCKS

7.3.3.1 Cultivars

The majority of plantations include Eurojapanese hybrids developed in France by INRA (*Institut National de la Recherche Agronomique*).[17] The cultivars have medium vigour, precocious and regular yields, and are suitable for medium- or high-density plantings. The nuts are large and with a good appearance, but their flavor is generally poor. Early-ripening hybrids are 'Bouche de Bétizac', 'Précoce Migoule', 'Bournette, Marlhac', 'Marsol', 'Maridonne', 'Vignols', 'Marigoule', and 'Maraval.' The cultivars 'Maraval', 'Marigoule', 'Marsol', and 'Précoce Migoule' are used both on their own roots and as rootstocks for superior varieties. In California, common hybrids are 'Colossal', 'Nevada', and 'Eurobella.' The Eurojapanese hybrids 'Primato' and 'Lusenta' were released in Italy but are no longer marketed.

7.3.3.2 Rootstocks

Popular clonal rootstocks for Eurojapanese hybrids include 'Marsol' (CA 07), 'Marigoule' (CA 15), 'Maraval' (CA 74), and 'Marlhac' (CA 118); 'Ferosacre' (CA 90) is sensitive to low winter and spring temperatures (Table 7.1).

7.3.4 ORCHARD MANAGEMENT

Among the Eurojapanese hybrids, good pollinizers and astaminate cultivars can be found (Figures 7.13 and 7.14). Cultivars like 'Bouche de Bétizac' have sterile male flowers. Eurojapanese hybrids have a higher demand for water than European varieties, and an annual rainfall of 1200–1300 mm/year (or proper irrigation) is necessary to ensure a sufficient water supply (Table 7.2). New high-density plantings are typically trained to a high open-center form.

TABLE 7.1

Main Eurojapanese Cultivars Developed in France

Clone and Cultivar	Genetic Origin	Double Fruits %	Use	Features	Catkins	Nuts/kg
CA 74 *Maraval*	Natural hybrid *C. crenata* × *C. sativa*	<12	Fresh; processed	Scion cultivar rootstock	Longistaminate	55
CA 15 *Marigoule*	Natural hybrid *C. crenata* × *C. sativa*	<12	Fresh	Scion cultivar; rootstock; wood producer	Longistaminate	55
CA 112 *Bournette*	Natural hybrid *C. crenata* × *C. sativa*	<12	Fresh; confectionery industry; canned	Medium vigor; early ripening; high productivity	Longistaminate	60
CA 48 *Précoce Migoule*	Natural hybrid *C. crenata* × *C. sativa*	>12	Canned	Medium/weak vigor; early ripening; high productivity	longistaminate	60
CA 07 *Marsol*	Natural hybrid *C. crenata* × *C. sativa*	>12	Canned	Scion cultivar; rootstock; wood producer	Longistaminate	60
CA 43 *Vignols*	Natural hybrid *C. crenata* × *C. sativa*	>12	Canned	Fruits production, pollinizer	Longistaminate	50
CA 125 *Bouche de Bétizac*	Controlled pollination *C. sativa* × *C. crenata* × C 04	<12	Fresh; processed	Scion cultivar (also grafted on *Marsol*); productive	Astaminate	60
CA 118 *Marlhac*	Controlled pollination *C. sativa* × *C. crenata* (cv *Laquepie* × *C. crenata*)	<12	Fresh	Rootstock; scion cultivar	Astaminate	55
CA 124 *Maridonne*	Controlled pollination *C. sativa* × *C. crenata* (cv *Sardonne* × C 04)	<5	Fresh; processed	Scion cultivar	Longistaminate (low fertility)	60
CA 122 *Marissard*	Controlled pollination *C. sativa* × *C. crenata* (cv *Laquepie* × *C. crenata*)	<12	Fresh; canned	Scion cultivar	Mesostaminate	60
CA 75 (*C. mollissima*)		<5		Pollinator with early flowering	Longistaminate	100

Source: Bounous, G., *Il castagno: Risorsa multifunzionale in Italia e nel mondo*, Edagricole, Bologna, Italy, p. 420, 2014.

FIGURE 7.13   'Précoce Migoule' is a good Eurojapanese hybrid pollinizer. (Courtesy of Gamba, G.)

FIGURE 7.14   'Bouche de Bétizac' orchard in northern Italy. (Courtesy of Beccaro, G.)

TABLE 7.2
Suggested Pollinators for Some Japanese and Eurojapanese Cultivars

Cultivar	Catkins	Pollinators
Ginyose (**J**)	Longistaminate	*Tsukuba* (J)
Tsukuba (**J**)	Longistaminate	*Ginyose* (J), *Vignols* (H)
Bouche de	Astaminate/	*Belle Epine* (E), *Marron de Chevanceaux* (E), *Marron*
Bétizac (**H**)	brachistaminate	*de Goujounac* (E)
		To a lesser extent: *Bournette* (H), *Maraval* (H), *Marsol*
		(H), *Précoce Migoule* (H)
Bournette (**H**)	Longistaminate	*Belle Epine* (E), *Maraval* (H), *Marigoule* (H), *Marsol*
	(medium-fertile pollen)	(H), *Précoce Migoule* (H)
Maraval (**H**)	Longistaminate	*Bournette* (H), *Marigoule* (H), *Précoce Migoule* (H)
	(low-fertile pollen)	
Maridonne (**H**)	Longistaminate	*Belle Epine* (E), *Marron de Goujounac* (E), *Verdale* (E)
	(low-fertile pollen)	
Marigoule (**H**)	Longistaminate	*Belle Epine* (E), *Marron de Chevancheaux* (E) *Marron*
	(medium-fertile pollen)	*de Goujounac* (E), *Portaloune* (E), *Bournette* (H),
		Maraval (H), *Marsol* (H), *Précoce Migoule* (H)
Marsol (**H**)	Longistaminate	*Belle Epine* (E), *Bournette* (H), *Marigoule* (H), *Précoce*
		Migoule (H)
Précoce	Longistaminate	*Belle Epine* (E), *Bournette* (H), *Ginyose* (J), *Marigoule*
Migoule (**H**),	(medium-fertile pollen)	(H), *Marsol* (H), *Tsukuba* (J), *Vignols* (H)
Vignols (**H**)	Longistaminate	*Ginyose* (J), *Marigoule* (H) *Précoce Migoule* (H)

Source: Bounous, G., *Il castagno: Risorsa multifunzionale in Italia e nel mondo*, Edagricole, Bologna, Italy, p. 420, 2014.

Note: (H) Eurojapanese hybrid; (J) Japanese chestnut; (E) European chestnut.

REFERENCES

1. Bounous, G. and Marinoni, D.T. 2005. Chestnut: Botany, horticulture, and utilization. *Horticultural Reviews* 31: 291–347.
2. Shen, G. 2015. *Chestnut-Science and Practice of Fruit Trees in China*. Taiyuan, Shanxi: Shanxi Science and Technology Press.
3. Guo, L., Hu, B., Dai, J.-H. and Xu, J.-C. 2014. Response of Chestnut flowering in Beijing to photosynthetically active radiation variation and change in recent fifty years. *Plant Diversity and Resources* 36(4): 523–532.
4. Pereira-Lorenzo, S. and Ramos-Cabrer, A. 2004. Chestnut, an ancient crop with future. In *Production Practices and Quality Assessment of Food Crops*, Volume 1, pp. 105–161. Dordrecht, the Netherlands: Springer.
5. Guo, S., Zou, F. and Xie, P. 2013. Comprehensive evaluation and screening of different pollination combinations of chestnut based on different evaluation methods. *Journal of Beijing Forestry University* 35(6): 42–47.
6. Zhang, X., Deyi, Y., Feng, Z., Xiaoming, F., Jing, T. and Zhoujun, Z. 2016. Studies on the pollen xenia of *Castanea henryi*. *Acta Horticulturae Scinica* 43(1): 61–70.
7. Anagnostakis, S.L. 2010. Chinese Chinquapin *Castanea henryi*. *Arnoldia* 68(2): 61–62.

8. Fan, X., Yuan, D., Tang, J., Tian, X., Zhang, L., Zou, F. and Tan, X. 2015. Sporogenesis and gametogenesis in Chinese chinquapin (*Castanea henryi* (Skam) Rehder & Wilson) and their systematic implications. *Trees* 29(6): 1713–1723.

9. Wang, G., Yuan, D.-Y., Zou, F., Xiong, H., Zhu, Z.-J., Liu, Z.-Q. and Ou, Y.-F. 2017. Effects of different pruning intensity on leaf physiology and yield in *Castanea henryi*. *Plant Physiology Journal* 53(2): 264–272.

10. Tian, S.-L., Sun, X.-L., Shen, G.-N. and Xu, L. 2015. Effects of compound fertilizer of $(NH_2)_2CO$ and KH_2PO_4 on the chestnut photosynthesis characteristics, growth and fruiting. *Chinese Journal of Applied Ecology* 26(3): 747–754.

11. Saito, T., Kotobuki, K., Sawamura, Y., Abe, K., Terai, O., Shoda, M., Takada, N., Sato, Y., Hirabayashi, T. and Sato, A. 2009. New Japanese chestnut cultivar 'Porotan'. *Bulletin of the National Institute of Fruit Tree Science* 9: 1–9.

12. Tanaka, K. and Kotobuki, K. 1992. Comparative ease of pellicle removal among Japanese chestnut (*Castanea crenata* Sieb. et Zucc.) and Chinese chestnut (*C. mollissima* Blume) and their hybrids. *Journal of the Japanese Society for Horticultural Science* 60(4): 811–819.

13. Takada, N., Sato, A., Sawamura, Y., Nishio, S. and Saito, T. 2010. Influence of pollen on pellicle removability and nut weight of Japanese chestnut (*Castanea crenata* Sieb. et Zucc.) 'Porotan'. *Acta Horticulturae* 866: 239–242.

14. Araki, H. 2004. *Chestnut Practical Handbook*. Tokyo, Japan: Noubunkyo.

15. Takeda, I. 1996. *Chestnut: Production, Processing and Selling*. Tokyo, Japan: Noubunkyo.

16. Sakamoto, D., Inoue, H., Kusaba, S., Sugiura, T. and Moriguchi, T. 2015. The effect of nitrogen supplementation by applying livestock waste compost on the freezing tolerance of Japanese chestnut. *The Horticulture Journal* MI-046.

17. Bounous, G. 2014. *Il castagno: Risorsa multifunzionale in Italia e nel mondo*. Edagricole, Bologna , Bologna, Italy, p. 420.

8 European Chestnut Traditional and High-Density Orchards

CONTENTS

8.1   TRADITIONAL PLANTATIONS RECOVERY

Gabriele Beccaro, Giancarlo Bounous, and Giulia Tessa

Traditional European chestnut groves represent an economic, environmental, historical, and cultural heritage.[1] Many plantations need renewal and recovery after years of abandonment or following attacks of pests and diseases that have compromised their efficiency.[2,3] The techniques must be calibrated according to the phytosanitary status, productive system, and orographic situation of the groves. In the worst orographic situations, the best solution is to convert the grove to a coppice. Recovery actions in areas where canker blight is virulent and in steep, easily erodible soils, or those heavy, humid, and cold areas, which retain water and promote ink disease, are unreliable. The groves to be restored must also be chosen according to the ease of access and viability, the presence of technically prepared growers, and the existence of infrastructure such as centres for technical assistance, processing, and marketing of the produce. It is also necessary to evaluate whether the existing varieties are valuable according to the chosen market destination (fresh market, processing, drying, flour).

If the pedoclimatic, orographic, and general conditions are suitable for chestnut production, the plantations may be recovered. The recovery of old traditional groves passes through a more or less severe pruning of the plants according to the level of decline of the plants. Generally, dry, old, and canker blight-affected branches have to be eliminated in order to stimulate the growth of a new and vigorous canopy. At the same time, chestnut groves must be cleaned to eliminate ferns, bushes, and weeds.

An increase of the organic matter in the soil can be gained through sheep and cattle pasture, outside the period of harvesting the fruits, or with the use of manure (10–15 t/ha/year). Mowed grass, leaves and husks, and the small pruned material (not infected) can be left on the ground, as they provide further enrichment in organic matter. Fertilization of the whole plot has to be carried out according to soil conditions (after soil analysis) and uptakes (see Chapter 6). For example, in an average mature European orchard, the annual fertilization could consist of 0.3–0.4 t/ha of ammonium nitrate and, when needed, 0.1–0.2 t/ha of mineral perphosphate and 0.2–0.3 t/ha of potassium sulphate.

8.1.1 CULTIVARS

The germplasm of *Castanea sativa* is rich in excellent cultivars grown in the main areas; Italy, France, and Spain, for example, have hundreds of varieties selected for candying, roasting, drying, and flour. Where the environmental conditions are very favourable, the best varieties are the Marrone type ('Chiusa Pesio', 'Val Susa', 'Castel del Rio', 'Marradi', and 'Fiorentino' in Italy; 'Montagne', 'Sardonne', and 'Comballe' in France). Italian Marroni, whose nuts are demanded by the market for their large size, and are appreciated both for the fresh market and for candying (marrons glacés), are also cultivated in Chile and Australia. Many varieties, with small but very sweet and easy-to-peel nuts, are suitable to be dried and for flour. Cultivars recommended in France for new plantations are 'Dorée de Lyon', 'Bouche Rouge', 'Belle Epine', and 'Marron de Goujounac'. The most important cultivars in Spain are 'Parede' and 'Longal'. In Portugal, 'Longal', one of oldest varieties, is widely spread over all the chestnut regions and has been promoted as the best cultivar for industry. 'Judía' and 'Martaínha', due to their larger nut size, are usually preferred for the fresh market. In California, 'Silverleaf' is widespread. See Chapter 4 for the complete cultivars list.

8.1.2 RECOVERY OF GRAFTED TREES WITH VALUABLE CULTIVARS

Depending on the age of the plants, the severity of the parasitic attacks, and the period of abandonment, it is possible to identify different actions. When the chestnut grove includes valuable varieties but the trees, senescent or sick, have a slow vegetative growth and reduced fruit size, a drastic pruning of the canopy is necessary.

The techniques consist of toppings and often large-diameter cuts, followed in the first 4–5 years by selection of new shoots to regulate the shape of the canopy. The pruning must be coupled with cleaning of the undergrowth and with the elimination of intrusive species. The cutting surfaces must be well defined and the wounds covered, when possible, with an anticryptogamic mixture (usually containing copper; Figure 8.1) and with cicatrizing dressings, even if not all the authors agree on the effectiveness of this last action.

The cuts must be executed with tools carefully disinfected with alcohol or fungicides to avoid the transmission of infections. It is advisable to restore entire groves and not to operate on single trees, to avoid damaging already pruned trees. The recovery must include fertilization, periodic cutting of the grass, and adjustment of the roads to facilitate the access to the chestnut grove.

FIGURE 8.1   Anticryptogamic mixture on cutting surface. (Courtesy of Beccaro, G.)

8.1.2.1 Light Pruning for Partially Compromised Chestnut Trees

Gradual and light renewal consists of eliminating the dry, senescent, sick parts and the overlapping branches and those that give the tree a sharp shape (Figure 8.2); lightening the central part of the canopy promotes the penetration of light and so, too, the production of flowers. For these trees, a regular pruning every 5–7 years is recommended (Technical Sheet 8.1).

FIGURE 8.2 Light pruning for partially compromised chestnut trees. (Courtesy of Beccaro, G.)

TECHNICAL SHEET 8.1 LIGHT AND SEVERE PRUNING FOR PARTIALLY COMPROMISED CHESTNUT TREES

1. Partially compromised chestnut tree

2. Light pruning of the canopy.

3. Severe pruning of the canopy with the elimination of the dry, senescent and sick parts.

Courtesy of Facello, V.

8.1.2.2   Severe Pruning of Heavily Decayed Chestnut Trees

In the case of old trees that are very compromised by canker blight, a large part of the canopy is removed, with drastic cuts to shorten the main branches (Figures 8.3a–c and 8.4). In following years, pruning allows regulation of the shape and the density of the canopy, so that from the fifth year it can be possible to obtain a regular harvest. The new year sprouts are selected to form, in 3–4 years, a balanced canopy, well distributed in the space, to best intercept the light. The basal suckers and each sprout on the rootstock must also be eliminated. A reduction in the height and volume of the canopy almost always attenuates the occurrence of cortical canker. The architecture of the tree should be generally respected, promoting the development of young

(a)

(b)

(c)

FIGURE 8.3 (a, b, c) Different examples of severe pruning. (a: Courtesy of Fabro, M.; b: Courtesy of Beccaro, G.; c: Courtesy of Gamba, G.)

FIGURE 8.4   Tree canopy after severe pruning. (Courtesy of Gamba, G.)

branches inserted above the grafting point, including them in the structure of the canopy. The height and lateral expansion of the canopy have to be reduced, and the emission of new shoots on the lower canopy has to be stimulated. Concerning monumental chestnut trees, pruning has as its primary objective the increase of mechanical stability and aims to preserve and enhance the grandeur of the trunk and the complex articulation of the canopy. Return cuts on the branches (never on the trunk) are needed (Technical Sheet 8.2).

TECHNICAL SHEET 8.2   SEVERE PRUNING OF HEAVILY DECAYED CHESTNUT TREES AND CANOPY EVOLUTION

1. Heavily decayed chestnut tree, but grafted with a valuable cultivar.

2. Removal of a large part of the canopy.

3. First year after the severe pruning.

4. Second year after the severe pruning, selection of the new spouts.

5. Third year after the severe pruning, selection of the new spouts.

6. 6- to 7-year-old chestnut tree after the pruning.

Courtesy of Facello, V.

The intensity of pruning, determined by the quantity of canopy removed, is one of the factors that determine its effectiveness: the best result is represented by a compromise between reinvigoration of the tree and the maintenance of a certain productive capacity.

Fertilization has always to be provided, considering uptakes and a soil analysis (see Chapter 6).

8.1.2.3 Topping Compromised Trees

Topping is an extreme technique that should be applied rarely, only when the architecture of the grafted tree is completely compromised. Branches derived from topping are weakly inserted into the trunk. The selection of new shoots must take place during the second and third winter after topping. The criteria for shoots selection are: (1) shoots orientation—eliminate shoots directed inwards (no more than 40° from the vertical); (2) if double shoots are present, eliminate one; (3) the morphology of the shoot collar should be strongly and correctly inserted into the trunk, with an angle not too open and not too closed; (4) the shoots presenting canker blight have to be eliminated as a priority. Moreover, it is appropriate to leave more, less vertical (45° angle) shoots on the side from which the prevailing wind comes (Technical Sheets 8.3 and 8.4).

TECHNICAL SHEET 8.3   TOPPING COMPROMISED TREES

1. Topping should respect, as much as possible, the tree architecture (primary branches).

2. A heavy topping on primary branches should be applied only in highly compromised trees.

3. Topping the chestnut tree on the trunk should always be avoided and only performed if the tree is grafted with a valuable cultivar.

4. New shoots development from the trunk after topping the trunk: the new branches are
weakly inserted on the tree. Tree architecture is compromised.

Courtesy of Facello, V.

TECHNICAL SHEET 8.4   TOPPING HIGH BRANCHED CHESTNUT TREES

1. Chestnut tree with too high branches.

2. Tree height reduction, maintaining tree architecture
and main branches structure.

3. New canopy formation.

Courtesy of Facello, V.

8.1.3 RE-GRAFTING OF HEAVILY COMPROMISED TREES AND GRAFTING COPPICES TO BE CONVERTED TO ORCHARDS

If the grafted variety is poor, when the plants are strongly decayed, or if it is necessary to convert a coppice to an orchard, trees can be cut off and trees from natural regeneration in the chestnut grove can be grafted or, in the following year, the shoots originating from the stumps can also be grafted (Technical Sheet 8.5) (Figure 8.5a and b).

TECHNICAL SHEET 8.5   RE-GRAFTING ON STUMPS FOR HEAVILY COMPROMISED TREES

1. Strongly decayed tree.

2. Cut off tree.

3. Year 1: shoots originating from the stumps.

4. Year 1: shoots selection.

5. In year 2 the shoots are grafted (3 to 4) and a new canopy is formed.

Courtesy of Facello, V.

The stumps should be chosen among those that produce strong shoots, placed at a regular distance from each other (about 10 m), to obtain an average density of 100–110 trees/ha, adequate to best exploit the available soil and light. For each stump, another two or three external shoots have to be selected, as they are able to more easily develop an independent root system with respect to the stump of origin, and they will be grafted with the chosen varieties. The type of grafting (crown, split, triangle, English double split, chip budding) depends on the size of the available rootstock and the grafting period. For diametric, triangle, and crown splitting, subjects of 2–4 years of age, with a diameter of 5–6 cm, grafted at 100–150 cm from the ground can be used (Technical Sheets 8.6 through 8.8) (Figure 8.6a–c).

(a)

(b)

FIGURE 8.5 Coppices converted to orchard: (a) topping the trees; (b) grafted trees after one year. (a, b: Courtesy of Beccaro, G.)

TECHNICAL SHEET 8.6 GRAFTING
SHOOTS FORM THE STUMP

1. 6–9 shoots of 1 year are selected, well inserted into the stump.

2. At the end of the first year, or in the second year, shoots are grafted.

Courtesy of Facello, V.

TECHNICAL SHEET 8.7   DECAYED TREES
GRAFTED WITH A VALUABLE CULTIVAR

1. Primary branches cut and crown-grafted.

Courtesy of Facello, V.

TECHNICAL SHEET 8.8 CROWN GRAFTING ON COPPICE

Courtesy of Facello, V.

One-year-old suckers of 1–1.5 cm diameter are best for English double split, chip budding, and flute splitting. The success of the grafts depends on the correct period of execution, graft affinity, shoots and scions morphology, and protection of the grafting point with a dressing to reduce both pathogen infections and dehydration of the grafts. To limit the risks of canker infections, particularly high at the point of grafting, inocula can be applied with hypovirulent strains (see Chapter 11).[4,5]

To avoid the breaking of young shoots by the wind, they must be shortened and tutored to poles or canes. The risk of breaking is particularly high for crown grafts. Winter or summer pruning should avoid the buds that originate on the rootstock overcoming the graft. Agronomic care must continue in the following years. To contain the development of weeds and to facilitate the harvesting of the fruits, cleaning of the undergrowth must be carried out by regular mowing and elimination of bushes, ferns, and other trees. Adequate protection of the grafted trees against wild animal attacks must also be provided, such as by fertilization. Regular monitoring (once each month from April to October) should be done to check re-growth and ensure soil maintenance.

8.1.4 TREE CLIMBING FOR THE RECOVERY OF CHESTNUT GROVES

Tree climbing is a technique that allows access to the chestnut tree canopy with the aid of ropes to perform pruning and to monitor the health of trees (Figure 8.7a and b).

FIGURE 8.6 Coppices converted to orchard: (a) grafting with two or three external shoots; (b) shoot fixing; (c) grafted stumps. (a–c: Courtesy of Beccaro, G.)

FIGURE 8.7   (a, b) Pruning with the tree climbing technique. (a: Courtesy of Beccaro, G.; b: Courtesy of Gamba, G.)

The technique allows movement from the inside to the outside of the canopy following the natural growth lines, allows to work in the canopy while pruning, and avoids damage to the chestnut as can happen with the use of other traditional working techniques. Moreover, it allows evaluation of the presence of diseases (e.g. cortical canker) and of structural problems of the trunk and the canopy that are difficult to see from the ground. Tree climbing allows work on trees cultivated in places not accessible in other ways, and it avoids compaction of the soil.

8.2   HIGH-DENSITY EUROPEAN CHESTNUT ORCHARDS

Pedro Halçartegaray Riqué

8.2.1   WHY HIGH DENSITY FOR EUROPEAN CHESTNUT?

When *C. sativa* grows naturally and in full sunlight, it is a very big tree, wide, and with a round shape, and with big limbs which give the tree that greatness that impresses anyone. It is beautiful in all seasons, in spring with its plentiful and scented yellow flowering, in autumn with its yellow leaves, in summer crowded with fruits, and in winter showing its magnificent structure (Figure 8.8).

But it is not an easy tree to domesticate for fruit production like other fruit tree species. In the wild, it is a quite efficient tree for wood production. This is because chestnut trees produce fruits only on the exterior of the canopy that is usually on the outer 50 cm that surround the canopy. In Table 8.1 the fruiting efficiency of a *C. sativa* tree with an open round shape of 12 m diameter with one with a conical shape with 7 m high and 5 m wide is compared (Figure 8.9a and b).

In the first case, the tree has spent lot of years and energy building a large amount of wood instead of fruits. But this is not the only reason to make a change in the chestnut canopy shape in high-density plantations. The other reason, equally important, is because in an orchard the round shape tends to prevent light penetration when trees become older and touch each other. A conical shape allows light to penetrate to the bottom part of the tree when the orchard is older, so the fruiting potential is not decreased by lack of light (Figure 8.10a and b).

It can be concluded that both the vegetative vigour and the shape must be controlled if the chestnut is to be an efficient fruiting tree in high-density orchards.

FIGURE 8.8   A *C. sativa* tree in its natural habitus. (Courtesy of Halçartegaray, P.)

TABLE 8.1

Comparison Between the Fruiting Efficiency of a *C. sativa* Tree with an Open Round Shape of 12 m Diameter with One with a Conical Shape with 7 m High and 5 m Wide, Considering a 0.5 m Productive Branches Length

Tree Canopy Shape	Potential Fruiting Volume	% Productive Volume
Round shape with 12 m diameter	$V = 4/3 \times \Pi \times r^3$	23%
Conical shape with 7 m high and 5 m wide	$V = \Pi/3 \times r^2 \times h$	45%

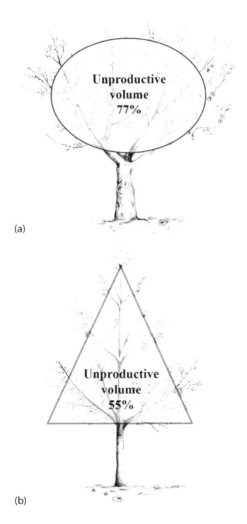

(a)

(b)

FIGURE 8.9   Unproductive volume of: (a) *C. sativa* tree with an open round shape with 12 m diameter; (b) *C. sativa* tree with a conical shape with 7 m high and 5 m. (a, b: Courtesy of Halçartegaray, P. et al.)

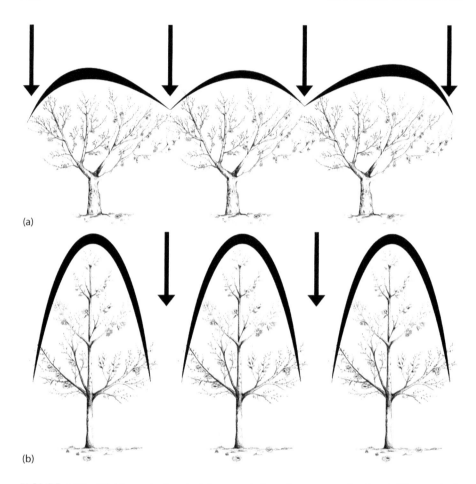

FIGURE 8.10   Light penetration in (a) a round-shaped canopy orchard and (b) a conical-shaped canopy orchard. (a, b: Courtesy of Halçartegaray, P. et al.)

A quickly profitable harvest is the main criterion that should be met for a chestnut tree plantation. This happens when the allocated space in the orchard is filled with each tree, and that happens early when the allocated space is small (Figure 8.11a and b). The problem is that if the space is too small for the final tree size, very soon there will be lack of light and therefore the loss of an important part of the fruiting wood. This has happened frequently with chestnut when an attempt has been made to increase the density without controlling the vigor and shape (Figure 8.12). To work on those aspects, it is necessary to know well the behaviour of *C. sativa*, how it reacts to what is done to the tree, and which are the physiological mechanisms used to produce wood, flowers, and fruits.

> *Determining European chestnut tree vigour*: in young chestnut trees, all the physiological behaviours that stimulate vigour will retard fruit production. In this sense, high density has an advantage because it is not necessary to

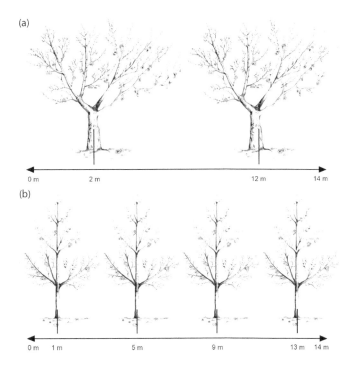

FIGURE 8.11 Different techniques of space allocation in an orchard: (a) it takes 15 years to fill the space; (b) it takes 6 years to fill the space. (a, b: Courtesy of Halçartegaray, P. et al.)

FIGURE 8.12 Chestnut trees grown with inadequate space allocation. (Courtesy of Halçartegaray, P.)

promote growth to fill a wide space, and this helps the tree to begin loading fruit earlier and more efficiently. It is necessary to consider four aspects of the behaviour of the *C. sativa* tree to reach our objective.

1. It is well established that 'vigour generates more vigour', what is called the vigour spiral, which can be observed in several deciduous trees. In Figure 8.13a a horizontal fruiting walnut wood with less vigor upon which it grew one vigorous vertical shoot is shown. This shoot has forced the limb to thicken down, and the trunk and some roots are connected to this limb. In Figure 8.13b a young chestnut with a central leader upon which three vertical vigorous shoots have been allowed to grow is shown; they have forced the trunk to thicken down, triggering the vigor spiral.

2. Trunk and root growth depends on the branches' demand. Only one vertical and vigorous branch is enough to trigger the vigour spiral. If only weak or horizontal low-vigor branches are allowed to grow, the whole tree tends to decrease its growth. On a limb, the vertical position increases vigor, and that promotes wood production. A horizontal or hanging position decreases vigor and promotes flowering.

3. Vertical shoots born from the leader have a narrow insertion angle and this results in deficient unions which tends to break (Figure 8.14a and b). Low-vigor horizontal shoots have wide insertion angles with solid crotches (Figure 8.15).

4. *C. sativa* only produces female flowers from the first to fifth terminal shoot on 1-year-old wood. But this occurs only in short shoots, less than 40 cm. Vigorous shoots that do not stop growing in the firsts weeks after bud break rarely produce female flowers.

Canopy shape. The big and round natural *C. sativa* canopy shape has its origin in a great dominance of basal branches, at the detriment of the central leader. This begins in the shoot apex of the new plant, where a great concentration of buds can be found (10 or more) in the top 10 cm (Figure 8.16a and b). In spring, these buds break forming a chandelier type, and gradually the lower shoots begin to dominate the terminal shoot, the one which should form the central leader. This works against the formation of a central leader. If there is no pruning, the same will happen the following year, forming another chandelier farther up, and basal dominance will be installed (Figure 8.17a and b).

In this way, a consistent amount of structural wood will be produced to support a low quantity of fruit on the external part of the canopy. The tree structure has to be minimized in order to have a more efficient tree for chestnut production. The simplest structure is a single central leader with hanging fruiting wood, without scaffold branches (Figure 8.18). With this kind of tree, if the vigor is under control, it is possible to fill the space and to crop early.

FIGURE 8.13 Different examples of vigor spiral: (a) horizontal fruiting walnut wood with less vigour upon which it grew one vigorous vertical shoot; (b) young chestnut with central leader upon which there are three vertical vigorous shoots. (a, b: Courtesy of Halçartegaray, P.)

(a)   (b)

FIGURE 8.14 Vertical shoots born from the leader weak (a) and broken (b). (a, b: Courtesy of Halçartegaray, P.)

FIGURE 8.15 Low vigor horizontal shoots with wide insertion angles and solid crotches. (Courtesy of Halçartegaray, P.)

(a)   (b)

FIGURE 8.16   (a) Concentration of buds in the shoot apex of the new plant; (b) buds developed after some weeks. (a, b: Courtesy of Halçartegaray, P.)

(a)   (b)

FIGURE 8.17   (a, b) Chandelier type. (a, b: Courtesy of Halçartegaray, P.)

FIGURE 8.18   Structure with a single central leader. (Courtesy of Halçartegaray, P.)

8.2.2   The Chilean High-Density Model

The Chilean high-density orchards are realized with the Italian Marroni and chest-nut varieties ('Marrone di Marradi', 'Marrone di Chiusa Pesio', 'Marrone di Susa', 'Marrone di Città di Castello', 'Marrone di Castel del Rio', 'Marrone di Castel Borrello', 'Garrone nero') and were planned and established at the beginning of the millennium in Chile by Dr. Pedro Halçartegaray. The chestnut high-density expe-rience in Chile was done under particular conditions. In the region where chest-nut is abundant, there are very deep volcanic soils with very high water retention but at the same time high porosity and fast infiltration. This could be strange, but these soils formed by volcanic microporous ash retain water inside their particles, not between them. So, the water that does not enter the pores drains fast, maintain-ing a high oxygen presence in the soil. This promotes good root growth in chest-nut and very good tree vigour. The soils have a ph of 5.5 to 7, with 10% organic matter and with enough nutrients to promote good chestnut production without the need for any fertilizers. Clonal rootstocks, with less vigour, are not diffused; *C. sativa* seedlings are used as rootstocks, and very vigorous Italian varieties are grown. The climate is very favourable for chestnut; rainfall is between 1.000 and 1.800 mm per year with only two drought months in summer and with a warm

FIGURE 8.19 High biodiversity rate in an orchard managed without pesticides. (Courtesy of Halçartegaray, P.)

summer temperature between 28°C and 30°C. All those conditions have made the control of vigour and shape of the trees more difficult. Our goal is to get high-crop orchards but taking care of biodiversity, and this is possible in our case due to the absence of pests that allows us to work without pesticides (Figure 8.19).

8.2.2.1 Planning the Orchard

The most important thing to be considered is soil selection. For *C. sativa*, as for the other *Castanea* spp., volcanic soils or sandy-loam soils with fast infiltration are preferred, never clay, compacted, or wet soils. Ideally, it is better to plant on slopes because flat soils tend to retain rain water for longer, causing asphyxia. On slopes, where possible, it is better not to plough the entire surface, to prevent erosion. If there are not heavy soil problems, only a 1.5 m band of soil should be ploughed, applying a basic fertilization if necessary. In the case of compacted soils, this band can be prepared with deeper work with an excavator equipped with a trident that penetrates 1 or 1.5 m deep, breaking and mixing the soil (Figure 8.20).

Plantation scheme. A rectangle is preferred, 7 × 5 m, with the larger distance between rows to allow better lighting and easier mechanization. Rows should be oriented from north to south for good lighting of both sides of the row, if the site allows it.

Pollinizers distribution. The Italian marrone type varieties planted in the high-density orchards in Chile all have astaminate catkins, so pollinizers are needed. Four marrone type rows and one pollinizer row should be planted. One early-flowering variety and one later pollinizer can be alternated. With this scheme, a good fruit set of 2.5 fruits per burr is usually obtained.

FIGURE 8.20   Excavator equipped with a trident. (Courtesy of Halçartegaray, P.)

FIGURE 8.21   Roots placement respecting its shape. (Courtesy of Halçartegaray, P.)

The pollinizers used in Chile were selected from seedlings and they produce good chestnuts, but not marrone type varieties.

Planting the trees. It is very important to place the roots correctly under an
horizontal radial profile, without twisting or rolling them, respecting the
shape they had in the nursery (Figure 8.21).

When planting, the crown must be 5 cm above the soil level to keep this part dry
to prevent *Phytophthora* infections. Regarding fertilization, depending also on
soil analysis, in the Chilean pedologic conditions, a little nitrogen (no more than

20 units/ha) has to be applied, usually only the first year of plantation; in these conditions, this is enough to get maximum growth of the leader of no more than 2 m. Higher growth is not advisable, because it can induce some bud malformation and makes training difficult. Some watering is recommended in the drought period in the middle of summer, depending on climate and soil conditions.

8.2.2.2 Training for High Density with a Central Leader

First year. Bare root trees, 1.8 to 2.2 m high, are planted. At planting time, the central leader has to be cut back to 50% of the annual growth, preferably choosing a bud oriented to the main summer wind. With this type of pruning, vigorous shoots will form a dominant central leader. In spring, the best shoot against wind will be chosen and the rest removed. The goal is to allow only one vigorous leader, 0.8 to 2.0 m high, in the first summer (Figure 8.22a and b).

Second year. In winter, pruning is not performed, except for the weak trees; in these trees, a third of the current growth is cut back to enhance vigour. In spring, the first year's work is repeated, growing one single leader. A 4 m bamboo stake can be used to guide the leader vertical. In the case of trees more than 1.6 m high not pruned in winter, in early spring the apical shoot that will continue the leader is selected; the other shoots are removed on the first 30–40 cm to control the competition that would form a chandelier structure (Figure 8.23a–c). The goal for the summer it is to get a 3 to 4 m-high central leader with some twigs grown on the last year's wood (Figure 8.24a and b).

(a)   (b)

FIGURE 8.22 (a, b) One vigorous leader of 0.8 to 2.0 m high. (a, b: Courtesy of Halçartegaray, P.)

(a)   (b)   (c)

FIGURE 8.23 (a) Selection of an apical shoot that will continue the leader; (b, c) shoot removal on the first 30–40 cm. (a–c: Courtesy of Halçartegaray, P.)

(a)   (b)

FIGURE 8.24 (a, b) High central leader with some twigs grown in the last year's wood. (a, b: Courtesy of Halçartegaray, P.)

In volcanic soils, fertilizers are not necessary. Excess vigour it is not convenient as this would request to remove the too vigorous twigs.

Third year. In winter, twigs are selected that will form the fruiting wood. They are weak, flexible, and tend to bend, getting a horizontal position. The vigorous vertical twigs must be removed. In *C. sativa* weak twigs are those having the following diameter ratio with the trunk:

$$a/b = {<}0.4$$

where a is the twig diameter and b is the trunk diameter. The diameter of the trunk is measured above the twig insertion point. A ruler may allow measurement of this relation (Figure 8.25a–c). Twigs that do not comply with the ruler or that have very narrow insertion angles should be removed (Figure 8.26a and b).

(a)

(b)   (c)

FIGURE 8.25 (a) Ruler for twigs selection; (b, c) weak twigs selected with a determined diameter ratio with the trunk. (a–c: Courtesy of Halçartegaray, P.)

(a)   (b)

FIGURE 8.26 Removal of twigs that do not comply with the ruler or that have very narrow insertion angles before (a) and after (b) the pruning in winter. (a, b: Courtesy of Halçartegaray, P.)

The pruning cuts must be horizontal. In this way, the upper buds that will give vertical shoots are eliminated and the lateral and lower buds that will give horizontal shoots are selected (Figure 8.27). Sometimes, there are buds on the leader that do not break, leaving some leader sections naked. In this case, in winter some incisions are made above these buds and this forces bud break in spring. This can be done on 2- and 3-year-old wood (Figure 8.28). In spring, if the leader has not reached 3 or 4 m high, last spring procedure can be repeated, that is to clean the first 30–40 cm from the top to eliminate the competition. If the leader has reached that height, nothing has to be done; it is only necessary to wait for the twigs to become fruiting wood, forming that typical terminal structure called 'chicken foot' (Figure 8.29a and b). This structure contributes with its weight to bend the branch, getting a horizontal position. This will be the fruiting structure that could be productive for many years.

FIGURE 8.27   Horizontal pruning cut. (Courtesy of Halçartegaray, P.)

FIGURE 8.28   Bud break in spring after incision in winter. (Courtesy of Halçartegaray, P.)

(a)   (b)

FIGURE 8.29 Fruiting structure called 'chicken foot' in (a) spring and (b) winter. (a, b: Courtesy of Halçartegaray, P.)

8.2.2.3   Fourth Year and Following Years

At the end of the third year, the basic structure of the central leader tree is formed, with its first fruiting limbs hanging at the bottom, and it begins to produce the new fruiting wood in the upper part; these last branches will be less vigorous and more productive than the bottom limbs in the young trees (Figure 8.30a and b). The more vigorous basal limbs will tend to crop less fruit in the following years. In the fourth year winter and in the following years, the following limbs always have to be eliminated:

- Limbs with an a/b diameter ratio up to 0.4;
- Vertical and strong branches that are not suitable to bend in the summer to a horizontal position with their own weight; their flexibility can be evaluated by the pruner when trying to bend it with their arm;
- Branches inserted on a weak crotch or with a bad insertion angle: usually this happens with vertical shoots.

In the tree shown in Figure 8.30b, one 'chicken foot' was left whilst four were pruned, giving rise to new twigs in their place. All new shoots must not be touched; it is not necessary to thin them because they bend themselves to a horizontal position. They will become fruit-bearing wood with a good insertion angle. Only limbs that are too vigorous and/or with a bad insertion angle are to be eliminated. Several buds are maintained where the limbs are cut off that will become new shoots with a good diameter with respect to the leader and good-quality fruiting wood (Figure 8.31).

If the job is properly carried out in the first 4 years, it will not be necessary to prune in the following years. The usual trend of unskilled pruners is to not cut some branches because of doubtful vigour or verticality, and this will cause problems, delaying the potential first harvest by triggering the spiral of vigour.

(a)   (b)

FIGURE 8.30   Basic structure of the central leader tree in (a) spring and (b) in winter. (a, b: Courtesy of Halçartegaray, P.)

FIGURE 8.31   New shoots with a good diameter from buds left where the limbs are cut off. (Courtesy of Halçartegaray, P.)

With the training system described, the potential crop can be very high, getting more than 30 kg per tree at year 7. In an orchard planted at 7 × 5 m, this means more than 7 t/ha (Figure 8.32a and b).

After year 10. After year 10, it may be necessary to do some renewal pruning. The branches begin to hang down and the upper fruiting centres of that branches

(a)   (b)

FIGURE 8.32   (a, b) High productivity of chestnut trees. (a, b: Courtesy of Halçartegaray, P.)

begin to dominate the lower ones. When these lower centres become weak and shady, they can be eliminated to renew the old fruiting wood.

8.2.2.4 Problems and Solutions

Before getting the full productive potential of the orchard, many problems can occur. The main problems are related to excessive tree vigour that delays production. In fact, if the training techniques fail, the tree will keep its juvenility, giving priority to vegetative growth to the detriment of flowering. If must be considered that vigorous shoots longer than 40 cm rarely produce female flowers on young trees.

The conditions that stimulate vigour in young chestnut trees are the following:

- Deep fertile soils with fast water infiltration;
- Vigorous rootstocks like *C. sativa* seedlings or clonals like 'Marigoule' or 'Marsol'. It will be better to try with 'Maraval' or others with less vigour;
- Fertilization in very fertile soils. Fertilization is necessary on shallow or poor soils only. In any case, it is better to retard fertilization until post-flowering;
- High irrigation or rains in spring, which enhances growth;
- Permanence of vigorous and vertical branches that trigger the spiral of vigour;

- Winter pruning of big branches on trees more than 5 years old will return juvenility to the tree. These branches must be eliminated until the 4th year;
- Sometimes it can be difficult to form this type of tree with some Eurojapanese hybrids, in particular with those that produce very vertical limbs with narrow angle insertions.

8.2.2.5 Reversing Vigorous Trees

When control of vigour on 5- or 6-year-old trees is lost and there is little flowering, a physiological tool has to be used to reverse this situation, that is, scoring the chestnut trunk in spring, at bud-break. A single cut must be done, with a knife, cutting all the bark until exposing the wood. In 20 days, the wound will be healed, and the conductive vessels reconnected (Figure 8.33a–c).

Scoring accumulates hormones and nutrients produced in the young shoots in the upper part of the tree, inhibiting terminal shoot growth with a consequent greater production of female flowers (Figure 8.34). This should not be a common practice; it is a 'lifejacket' to be used only in cases when control of vigour has been lost.

In Figure 8.35 on the right row, a 6-year-old orchard with excess vigour and poor fruiting is shown. The row on the left was scored. Scoring on young trees can

(a)

(b)

(c)

FIGURE 8.33 (a) Single cut in all the bark until wood; (b) wound healed and conductive vessels reconnected after 20 days; (c) knife used for the operation. (a–c: Courtesy of Halçartegaray, P.)

FIGURE 8.34 Inhibited terminal shoot growth with consequent production of female flowers. (Courtesy of Halçartegaray, P.)

FIGURE 8.35 Orchard with a scored row on the left and a not-scored row on the right. (Courtesy of Halçartegaray, P.)

increase the crop from 2 t to more than 6 t per ha. This practice can be avoided if the job is carried out properly in the first four years and the soil and the rootstock are chosen correctly.

8.2.2.6 Asphyxia Problems in High-Density *C. sativa* Orchards

In 2010, a 60-hectare chestnut orchard was planted in the central south region in Chile. This was known to be a place where ancient and very big European chestnut trees, not grafted, were grown. The climate is very suitable to chestnut cultivation, with 1.300 mm rainfall, well distributed and with only two drought months in the middle of summer. The soil consists in volcanic ash in the first 1.5 to 2 m, with a fast infiltration rate, but in the lower profile there is a mix of ash and clay that slows infiltration. Half of the surface is on slopes and the other half is a flat terrace above 10 m above the river level. Italian marrone type varieties were planted, with local *C. sativa* pollinators and the hybrid pollinator 'Precoce Migoule'. All the trees were grafted on *C. sativa* seedlings in the same nursery. In 2015, in the 2 months before bud-break (October in the southern hemisphere), rainfall was concentrated, and 750 mm accumulated in that period. The soil in the flat area of the field was saturated for many days. A few days before bud-break, the symptoms of bubble bark, sour sap, or 'frog skin', began to appear on the bark of some trees. The bark was blackened and had blisters on its surface. A cut with knife revealed a dark sap and the smell of fermentation (Figure 8.36a and b).

At bud-break, some basal limbs did not start or started weakly, and some whole trees never began growing. But 1 or 2 months later, vigorous suckers began to grow on many trees at the crown or under the graft union. In a short time, these shoots were 1 or 2 meters high. In Figure 8.37 no damage can be seen on the slopes but

(a)   (b)

FIGURE 8.36   Symptom of bubble bark, sour sap or 'frog skin': (a) bark blackened and with blisters on its surface; (b) dark sap after a cut. (a, b: Courtesy of Halçartegaray, P.)

FIGURE 8.37   Damage on the flat area. (Courtesy of Halçartegaray, P.)

FIGURE 8.38   Affected trees with low production. (Courtesy of Halçartegaray, P.)

a devastating damage in the flat area. Some affected trees survived and produced many flowers but with a weak growth and many small chestnuts (Figure 8.38). In the next 2 years, these survivors had grown very poorly and produced many small chestnuts, so we decided to remove them. When cutting the trunk, it can be seen that the xylem formed before 2015 was dead. Only new xylem was active (Figure 8.39).

FIGURE 8.39   New active xylem and dead xylem before 2015. (Courtesy of Halçartegaray, P.)

One of the explanations that has been given is that a lack of oxygen in the soil produced fermentation of the reserve sugars in the roots and that produced some toxins like alcohols or lactic acid. These substances went up through the xylem, killing its cells, but the roots remained alive. In the case of that orchard, due to the large number of damaged trees, it was possible to obtain some interesting information about varietal susceptibility. The trees' mortality was: 57.7% for 'Marrone di Chiusa Pesio', 55.4% for 'Marrone di Marradi', 51.1% for 'Marrone di Susa', 39.9% for 'Marrone Città di Castello', and 6.4% for 'Precoce Migoule'. The events observed make us think that susceptibility to a lack of oxygen in the soil depends more on the scion (species and cultivar) rather than on the rootstock. It would be important to explore more deeply many more aspects: which processes occur in the root when oxygen is absent; which are the substances produced that are able to kill the xylem cells; why the xylem of some varieties is more sensitive than others; and what could be the relationship with the population of some microorganisms present in the roots. Perhaps the answers to these questions could allow us to expand the range of soils where chestnut could be grown in the future.

REFERENCES

1. Bounous, G. and Marinoni, D.T. 2005. Chestnut: Botany, horticulture, and utilization. *Horticultural Reviews* 31: 291–347.
2. Beccaro, G.L., Mellano, M.G., Barrel, A. and Trasino, C. 2009. Restoration of old and abandoned chestnut plantations in Northern Italy. *Acta Horticultuare* 815: 185–190.
3. Bounous, G. and Beccaro, G.L. 2002. Chestnut culture: Directions for establishing new orchards. *FAO-CIHEAM, Nucis Newsletter* 11: 30–34.

4. Bisiach, M., De Martino, A., Intropido, M. and Molinari, M. 1991. Nuove esperienze di protezione biologica contro il cancro della corteccia del castagno. *Frutticoltura* 12: 55–58.
5. Bounous, G. 2014. *Il castagno: risorsa multifunzionale in Italia e nel mondo.* Edagricole, Bologna, p. 420.

9 Postharvest Handling

Fabio Mencarelli and Andrea Vannini

CONTENTS

9.1 CHESTNUT CHARACTERISTICS AFFECTING POSTHARVEST BEHAVIOUR

Chestnut is a starchy nut with a cream-coloured cotyledon covered with a very thin membranc called pellicle (episperm). The ease of episperm removal is a required and appreciated characteristic for fresh market and processing. Unlike most true nuts, the chestnut does not have a stony, hard involucre surrounding the seed. The external, brown-black or brown-red involucre (pericarp) is very porous and facilitates water absorption but also permits rapid water loss from the seed.

Chestnut is almost entirely a seed but, in contrast to other nuts such as walnut, hazelnut, and almond, it is rich in starch (40%) and low in lipids (3%–4%). Moreover, chestnuts are rich in sugars, mainly monosaccharides and disaccharides such as sucrose, glucose, and fructose, and lesser amounts of stachyose and raffinose. Compared to dry nuts, the high-water content (around 50%), added to their high sugar content, makes chestnuts prone to sprouting and perishing due to water loss and disease. For these reasons, chestnut postharvest behaviour differs from that of other commodities. By comparing the main features that determine the perishability

TABLE 9.1

Main Characteristics Determining Perishability of Fresh and Dry Fruits Compared with Chestnuts

Characteristic	Fresh Fruit	Dry Fruit	Chestnut
Living organism	Yes	No	Yes
Water content %	90–95	5–15	50–60
Physical structure	Soft	Hard	Slightly hard
Respiration	High	No	Low
Transpiration	High	No	High

of a plant commodity, the differences among chestnut, fresh fruit, and dry nuts can be noted in Table 9.1.

The soft, leathery pericarp is a problem during mechanical harvest of chestnuts because it makes them susceptible to physical injury. The chestnut apex, or torch, is often removed during mechanical harvesting and handling practices, providing the main route of entry for fungal spores especially during the water curing process. The flesh of the nuts is hard compared to other fresh fruits, but soft compared to the flesh of some almonds or hazelnuts. For example, the average firmness of a commercially-ripe fresh fruit is 3–8 kg/cm^2; for chestnuts it is 10–20 kg/cm^2. Compared to dry fruits such as walnut, pecan, hazelnut, and almond, fresh chestnuts have the same or greater firmness but, also, greater elasticity which makes them less crispy and tougher at the first bite.

9.2 RESPIRATION AND ETHYLENE PRODUCTION

Seeds are usually characterized by a low and constant respiration rate without any peak (a non-climacteric pattern). As reported in Table 9.2, chestnut's respiration rate, depending on the temperature, varies between 2 and 10 mL O_2/kg-h,

TABLE 9.2

Physiological Characteristics and Storage Requirements

(a) Physiological Characteristics

T°	Respiration (mL O_2/kg-h)	Ethylene (μL/kg-h)
0°C	2.5–5	0.001
20°C	8–10	0.01

(b) Storage Requirements

Temperature	RH	O_2	CO_2	Time
0°C ± 1°C	90%–95%	—	—	4 mo
—		2%–3%	15%–20%	5–6 mo

(c) Tolerance to

Chilling	High CO_2	Low O_2	Heat	Low RH
Yes	Yes	>1%	No	No

which is similar to that of potatoes and onions and not much lower than that of some apple and pear varieties. Chestnuts, like other seeds, produce very little ethylene, but ethylene can be a good tool to control chestnut sprouting. The effect of ethylene is known to reduce potato sprouting during storage at low temperatures.[1]

9.3 WATER LOSS

Transpiration is a process that the living plant cell uses to lower its temperature, triggered by vapour pressure deficit (VPD = tissueVP−environmentVP). VPD is affected by air relative humidity (RH) or water content of the commodity and air or product temperature. The greater the VPD, the faster the transpiration rate. The higher the RH in the environment, the lower the VPD when the tissue and the surrounding environment are at the same temperature. If the temperature of the environment decreases, even maintaining the same RH, VPD increases until the fruit reaches the same temperature as the environment. When tissue and environment are at the same lower temperature, VPD is much lower than at higher temperatures.

Chestnuts have a high transpiration rate because the pericarp has no cuticle or other waxy materials to protect the seed against transpiration (or water penetration). In one day at 20°C and 70% RH chestnuts can lose as much as 1% of their weight. But if the temperature decreases rapidly and is kept low, VPD is reduced, as well as weight loss (Table 9.3). Simply reducing the temperature (while maintaining the same RH) reduces the weight loss by 3.5 times (in one day if chestnut loses 1% at 20°C and 70% RH, at 0°C and 70% RH will lose 0.3%); increasing RH to 90%, at the same temperature, will further decrease the loss (only by 0.17%).

Thus, temperature reduction and RH increase can ensure against a tenfold loss of profit. When there are mechanical injuries, even slight rupture or scratch of the pericarp or breakage of the torch, or in the case of pericarp darkening, water loss increases.

TABLE 9.3

Water Loss of Chestnut Fruits Expressed as VPD, Weight Loss, and Chestnut Loss

Temperature °C	RH %	VPD mbar	Weight Loss %/day	Chestnut Loss kg/t/day
20	70	7	1.00	10
0	70	2	0.30	3
0	90	0.6	0.17	1

9.4   SPROUTING

Chestnuts sprout rapidly in environmental conditions with high RH and high temperature. If chestnuts are kept at low temperature, sprouting occurs at the end of the winter, and sometimes earlier. Sprouting is favoured by water dipping of chestnuts. When sprouting starts, respiration increases as well as heat release, and these are difficult to control. High CO_2 and low O_2 have been shown to delay sprouting while water promotes it.[2] Continuous exposure of potato to chlorine for 40 days was shown to significantly reduce sprouting.[3] For this reason, the use of chlorine in water solution could control chestnut sprouting.

9.5   PESTS AND DISEASES

Damage to chestnut fruits is mainly caused by insects and fungi. Attacks by insects take place on the tree and can cause losses during storage and distribution. Among insects, the most relevant are chestnut moths (*Cydia fagiglandana* Zehl. and *Cydia splendana* Hb.) and chestnut weevils (*Curculio* spp.). Sorting out of infested fruits can be done partially during manual harvesting by discarding fruits with an evident adult exit hole. The introduction of mechanical harvesting makes sorting out before delivery of the product to the storage plant problematic. Furthermore, many larvae exit the infested fruits during the phases of storage and distribution. Removal of infested fruits is normally carried out by rapid passage of the fruits in cold water tanks. Infested fruits commonly float on the water surface and are removed mechanically or manually. The most common agents of superficial and inner contamination, and producers of secondary toxic metabolites (mycotoxins), belong to the genera *Penicillium* and *Aspergillus*. *Gnomoniopsis castanea*, the agent of brown rot (Figure 9.1), is recognized nowadays as the main cause of internal decay of chestnuts worldwide.[4]

FIGURE 9.1   Symptoms of brown rot caused by *G. castanea*. (Courtesy of Vannini, A.)

FIGURE 9.2   Symptoms of black rot of kernel caused by *S. pseudotuberosa*. (Courtesy of Vannini, A.)

Additional agents of decay are *Acrospeira mirabilis, Sclerotinia pseudotuberosa*, the agent of black rot (Figure 9.2), (syn. *Ciboria batschiana,* anamorph *Rhacodiella castanea*), *Amphiporthe castanea*, and *Fusarium* spp.[5] While fruit damage by insects is a direct consequence of infestation in the field, fruit contamination and decay by microorganisms (i.e. fungi) can be significantly amplified depending on postharvest fruit handling and storage.

Free water inside the chestnut, high RH, and room temperature are favourable conditions for fungus development. Depending on harvest season, postharvest treatments, and storage, losses by internal decay fungi (i.e. black and brown rot) can be 60%–70% of the final product.

Moreover, the presence of fungal contamination by *Aspergillus* spp. and *Penicillium* spp. is a food safety issue due to the potential for subsequent mycotoxin development. At present, assessment of the percentage of internal damage is carried out by random sorting and cutting of fruits during the product delivery phase at the treatment and storage plant. Manual removal of fruits with internal damage is carried out during the last phase of sorting on the conveyor belt, although this activity requires relevant expertise of the operators.

9.6   DISORDERS

The chestnut kernel is not normally susceptible to any specific disorder. It is not chill-sensitive, but it can freeze after harvest if the temperature is below −5°C for 1–2 days (temperature and time are inversely correlated). Similarly, chestnut is not CO_2-sensitive, but it can ferment in the absence of oxygen for several consecutive days depending on the presence of fermentable sugars in the seed but also on the occurrence of lactic bacteria and yeasts on the pericarp surface which can penetrate through pericarp injury and/or by maintaining chestnuts in water for postharvest treatment.

Non-infective internal browning is sometimes observed when cutting the seed at the end of storage period, and this discoloration is intensified by cooking. The colour alteration is due to reducing sugar accumulation (loss of starch due to hydrolysis) during senescence with cell disassembly.

A further non-infective disorder is the internal white chalky stain which appears in the seed during the storage. This stain is due to particular very small starch granules with a viscographic pattern showing a characteristic profile with high viscosity.[6] It is likely that, during storage and starch hydrolysis, in some parts of the seed, the water motion from inside to outside provokes difference in starch modification. Sprouting can be considered a disorder.

9.7   MECHANICAL INJURIES

In postharvest, three main mechanical injuries are coded: impact, abrasion, and compression. They are well described for all fruits but not for chestnut.[7] Apparently as aforementioned, chestnut seems a hard, strong product and, for this reason, its handling is very rough, and not high-technology. In contrast, chestnut is a product very susceptible to mechanical damage overall and to abrasion and bruises (impact). The first step where mechanical damage can occur is harvest, especially mechanical. In Table 9.4, the incidence of mechanical damage in chestnut comparing manual and mechanical harvested is shown.

These two types of injury can also occur during handling, especially during loading and unloading operations. Unfortunately, these injuries are not visible immediately (they do not cause a change in colour as for apple or pear) and they only become visible after water dipping and during storage or on cutting the chestnut. When the tissue is damaged (not visible), the rate of water absorption of the pericarp increases because of an increase in the absorbing surface. Immediately, this event facilitates fungus penetration and, successively, an internal decay is developed (Figure 9.3) or the pericarp colour changes to dark brown, due to greater oxidation of phenols.

Moreover, during water dipping, the detachment of the episperm from the pericarp is due to the formation of a water sac (Figure 9.4). In addition, another type of mechanical injury is breakage of the torch, which can become a fungus penetration point (Figure 9.5). In conclusion, chestnut handling should be revised by coding rules for loading and unloading and sorting operations.

TABLE 9.4
Percentage of Mechanical Injuries (Mean of Three-Year Harvest) Depending on Harvest Technique

	Cracks %	Bruises %	Abrasions %
Mechanical harvest	2.5	5.7	14.1
Manual harvest	1.0	1.2	4.5

Source: Cecchini, M. et al. *Postharvest Biol. Technol.* 61, 131–136, 2011.

FIGURE 9.3   Severe impact injury due to excessive height of drop in the loading and unloading operations. (Courtesy of Mencarelli, F.)

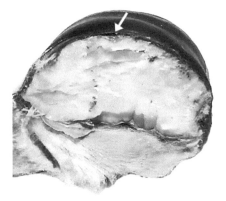

FIGURE 9.4   Light impact damage from 1 m height drop: the arrow indicates the episperm detachment. (Courtesy of Mencarelli, F.)

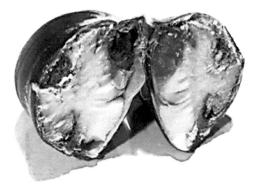

FIGURE 9.5   Torch break, mechanical injury, and decay development after one month of cold storage from the warm water treatment at the time of chestnut harvest. (Courtesy of Mencarelli, F.)

9.8   POSTHARVEST TECHNOLOGY OF CHESTNUTS

In postharvest handling of all crops, the '3Cs', cleanliness, care, and cooling, are important and even more so in chestnut. The reason is because chestnuts are harvested from the soil and most of the time they are dipped in water. In theory chestnuts should never touch water during postharvest treatments. In practice, water, cold or warm, is widely used in chestnut postharvest treatments with different aims, including restitution of the optimal water content, removal of infested fruits and debris, and sterilization from fungal contamination and tissue colonization. However, most of the time, such water treatments produce the opposite result, increasing the fungus inoculum and the risk of cross-contamination of fruits. This is related to the inappropriateness of the treatment protocols, and/or the employment of poor technology, rather than the difficulty in upscaling treatment protocols into the treatment plant.

9.8.1   Handling Procedures

Most of the steps today are gathered in a mechanical handling and, most of the time, automatized handling line where the fruit after several steps is packed for shipping. In the case of chestnut, handling has been always based on rough techniques because chestnut is still considered a dry, stony fruit, resistant to mechanical injuries. Today, the chestnut handling line is very similar to the potato one, with the difference that the price of chestnut in the market is far higher than that of potato. Mencarelli[8] described accurately the different approaches to chestnut handling, depending on the logistic destiny of chestnuts. A comparison between what is done today and what should be done in order to guarantee a high-quality and safe product to the consumer is presented below.

9.8.1.1   Traditional Mechanical Handling

1. Removal of the chestnut from the spiny burr (this practice is nowadays limited to areas with traditional hand picking of fruits, while it is abandoned where mechanical harvesting is employed)
2. Dry feeding: unloading chestnuts from the bin (usually 2.5 t approx.) into a tank and then, through a conveyor belt onto a perforated vibrating table to remove superficial materials together with forced air
3. Sizing by using a horizontal metallic perforated cylinder with holes of different sizes (Figures 9.6 and 9.7)
4. Grading: performed by visual assessment to sort out decayed chestnuts or foreign materials
5. Packing.

FIGURE 9.6   Rolling perforated cylinder for chestnut sizing. (Courtesy of Mencarelli, F.)

FIGURE 9.7   Sizing of chestnuts. (Courtesy of Mastrogregori Aldo, s.r.l.)

Critical points and advantages of actual, traditional mechanical handling are shown in Technical Sheet 9.1.

TECHNICAL SHEET 9.1   WEAKNESSES AND STRENGTHS OF TRADITIONAL MECHANICAL HANDLING AND INNOVATIVE SORTING

	Weaknesses	Strengths
Traditional mechanical handling	Dry feeding could produce mechanical injuries, depending on the height of unloading. Sizing by perforated screen is the worse in terms of size sorting because its quality depends on the amount (Q) of chestnuts loaded, the rotation speed, the cylinder inclination; the sizing depends only on the diameter; as chestnuts are not spheroidal or spherical, the perforated screen is very inaccurate. The efficiency of visual sorting depends on several factors: gender and age of the sorter, hours spent in grading operations, width of the grading table, light over the grading table, the Q on the grading table, temperature of packing house.	Low cost of machinery purchase and care. No computer controlling technologies.
Innovative automatic sorting	High sizing and grading performance. Number of operators.	Electricity source: a good electricity line must be guaranteed. Q must be quantified accurately to guarantee a good singulation. High cost of purchase. Accurate machine care and calibration.

9.8.1.2   Innovative Automatic Sorting

1. Dry feeding using an auto-adjusting bin dumper with continuous gradual and gentle fruit dumping such as is used for delicate fruits (Figures 9.8 through 9.10).
2. Brushing should be gentle to avoid torch breakage; accurate choice of brush hairs should be made (diameter, length, hardness).
3. Chestnut alignment and singulation: chestnut should be singulated in order to guarantee perfect sizing and grading.
4. Sizing by weight is the best way to guarantee an uniform lot of chestnuts for sale. In this case, the aforementioned $P = 40\%D + 60\%L$ where P is sizing parameter, D diameter, and L length, meaning that the machine is able to take into consideration both diameter and length and not only diameter as for the perforated screen.
5. Automatic grading for superficial defects and colour by machine vision.
6. Automatic box filling.

FIGURE 9.8 Gentle dry feeding avoids mechanical damage and torch break in chestnuts. (Courtesy of Unitec Group, S.p.A.)

FIGURE 9.9 Conveyor for chestnut singulation. (Courtesy of Unitec Group, S.p.A.)

FIGURE 9.10 Singulated chestnuts ready for sizing by weight and grading by vision system. (Courtesy of Unitec Group, S.p.A.)

Weaknesses and strengths of innovative automatic sorting are shown in Technical Sheet 9.1.

9.8.1.3   Innovation in Chestnut Sorting

In recent years, the rising impact of internal decay (brown and black rot), as well as the presence of larvae inside chestnuts, has pushed research towards the application of spectral techniques such as infrared spectroscopy and tomography. Moscetti et al.[9] tested NIR-AOTF to identify non-destructively the presence of insects inside chestnuts. The optimal spectral features corresponded to Abs [1582 nm], Abs [1900 nm], and Abs [1964 nm], and the results obtained represent an average of 55.3% improvement over a traditional flotation sorting system. Donis-González et al.[10] applied X-ray computed tomography (CT) on chestnuts to evaluate internal components and quality attributes. An image analysis method (algorithm) for the automatic classification of CT images obtained from 2,848 fresh chestnuts ('Colossal' and *C. mollissima* seedlings), during the harvesting years from 2009 to 2012, was developed and tested. The technique permitted internal, non-visible, decay identification with accuracy of more than 86%.

9.8.2   Postharvest Pre-storage or Pre-shipping Treatment

9.8.2.1   Water Curing

The main objective of this method is to reduce the potential development of fungi during the storage process and, at the same time, to favour the killing of insect larvae inside the fruits. Moreover, water curing is used for rough, inefficient, sorting of chestnuts affected by internal decay or worm holes because they float on the water surface. The efficiency of the method depends on partial lactic and alcoholic fermentation that takes place during the curing process, reducing pH, increasing ethanol and acetaldehyde content, and likely allowing the diffusion of phenols from the pericarp into the seed (Figure 9.11).[2,11] Today special stainless-steel tanks are used to keep the chestnuts immersed in water (Figure 9.12).

The following steps are important for a good response to water curing treatment:

1. Soak chestnuts in 15°C tap water for a maximum of 5 days (bubbling of water is an indication of fermentation taking place, developing CO_2).
2. Ratio of chestnuts: water = 1:1 (1 kg in 1 L).
3. After water curing, dry chestnuts very well by ventilation; the chestnut weight should be restored to the initial weight before water curing.

(a)

(b)

FIGURE 9.11   The color of water during chestnut curing depends on the ratio chestnut water: from the left to the right 2:1, 1:1, 1:2. Water at normal pH (a), water at pH 3 (b). (Courtesy of Mencarelli, F.)

FIGURE 9.12   Water curing tanks. (Courtesy of Mastrogregori Aldo, s.r.l.)

In Technical Sheet 9.2, strengths, weaknesses, and recommendations for water curing are shown.

TECHNICAL SHEET 9.2   WATER CURING AND HOT WATER TREATMENT: STRENGTHS, WEAKNESSES AND RECOMMENDATIONS

Treatment	Strengths	Weaknesses	Recommendations
Water curing	Enough fungistatic efficacy if all the recommendations are strictly applied. Increase chestnut weight. Partial sorting of culled chestnuts.	Huge amount of water; potential for water recovering by filtering and ozonation. Difficulty to guarantee the same fermentation condition in chestnut mass if water is still. Impossibility to remove all the free water from chestnuts. Favour chestnut sprouting. Chestnut drying is needed, energy-consuming. Presence of water facilitates fungi development in successive storage steps in absence of forced air drying of fruits.	Water must be drinkable or sanitized with ozone or NaOCl; chestnut must be moved in the water to avoid CO_2 accumulation; renew water each time new chestnuts are dipped for water curing. Clean the tank very well before starting the curing treatment. Dry chestnuts thoroughly before shipping but especially before storage.
Hot water treatment	Excellent sanitation power for internal decay agents: in controlled condition the survival of *G. castanea* at 50°C for 45 min is null (Figure 9.14). Sufficient sanitation power for external contaminants (*Penicillium*, *Aspergillus*). Increase of chestnut sweetness.	Relevant investment in equipment. High electricity consumption; high energy coefficient if an open water tank is used. Huge amount of water is needed; potential for water recovering by filtering and ozonation. Constant temperature of water and mass of fruits at each point of the tank is difficult to reach and maintain. In continuous processes the water in the tank gets dirty and contaminated: cleaning of re-circulating water is extremely complicated. In fact, ozone is difficult to use due to the high temperature of water (50°C); filtering is possible, but it increases substantially the amount of water needed for the treatment.	Chestnut must be submerged at least 20 cm below water surface, using a screen to avoid floating. Water must circulate constantly. Water must be kept at least at 50°C; the time of treatment starts when chestnut pulp has reached the temperature of water. Time of treatment depends on chestnut size. Check constantly the water temperature in different parts of the tank.

9.8.2.2 Fumigation

Quarantine treatment is a widely used treatment to control pests when a product is exported from continent to continent (more information on quarantine can be found in the USDA Manual, 2017). Since methyl bromide is nowadays forbidden for chestnut quarantine treatment, phosphine (Al phosphide or Mg phosphide) is widely used for chestnut fumigation because it is a simple, cost-effective method to remove stored product insect pests (SPIs) such as beetles, weevils, mites, and moths in bulk commodities such as chestnuts. Sulphuryl fluoride (ProFume®) is the active ingredient of a fumigant gas which comes in two forms.

Each country determines which gas can be used according to specific regulations and label requirements. Irradiation at 0.5 kGy has been recommended too, above all in the past, as one of the alternatives to MeBr fumigation for both quarantine and sprout control purposes.[12]

9.8.2.3 Hot Water Treatment

Hot water immersion, vapour heat, forced hot air, cold treatment, and irradiation are the most commonly used treatments as an alternative to chemical treatment for insect control. In chestnut, hot water treatment is performed before storage but mainly before shipping. It consists of dipping chestnuts in water at high temperature for a short time with the main objective to kill insects (worms) (Figures 9.13 and 9.14).

The main steps of the treatment are:

- Before hot water treatment, it is recommended to carry out a rapid immersion (a few minutes) in cold tap water to remove all the floating fruits from the mass;
- Water at 50–52°C for 30–45 min (higher temperatures up to 90°C for a few minutes have been proposed[13] but the cost of energy limits this application);
- Rapid cooling in cold tap water (30 min max);
- Draining of chestnuts and drying by forced air flow (10 to 15 min).

FIGURE 9.13 Hot water treatment chain: (1) cold tap water treatment and removal of floating fruits (few minutes); (2) hot water treatment tank at 50°C for 45 min; (3) cooling tank with cold tap water (max 30 min); (4) drying tank with forced air (ca 15–20 min). (Courtesy of Mastrogregori Aldo, s.r.l.)

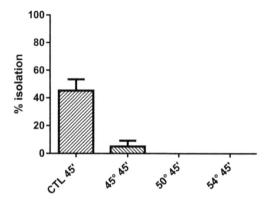

FIGURE 9.14   Efficacy of treatment in hot water on re-isolation on potato dextrose agar medium of the brown rot fungus *G. castanea*. Fruits were treated at 45°C, 50°C and 54°C for 45 min, followed by immersion in cold water for 60 min, dried to the original weight, and kept in a storage room at 5°C for 10 days. Controls received the same treatments except for hot water.

9.8.2.4   Other Treatments

To sanitize chestnuts, several additives to water can be used. As pointed out by Donis-González et al.,[14] the best results are obtained using Storox™, ozone (Aqua Air Techn. Inc), or ClO_2 (by ICA TriNova, Forest Park, GA). Storox™ and ozone gas in water are the most applied treatments, while ClO_2 is very risky to use. It is explosive at concentrations above 10% or at temperatures above 130°C but also when it comes into contact with ammonia compounds; it must be produced on-site, thus specialized workers are needed and a worker safety programme is requested, making this gas relatively expensive for produce application.

Storox™ is a strong oxidizer made up of hydrogen peroxide and acetic acid; it is very soluble in water with very little off-gassing, and it leaves no known toxic breakdown products or residue on the produce. It has good stability with organic matter, thus increasing the longevity of this sanitizer even with not very clean water, and it is not corrosive to equipment. Water must be acidified to a pH of around 5 in order to have better action, thus citric acid can be used before using Storox™. It cannot be used in hot water treatment because temperature reduces the sanitizing activity.

Organic acids (sorbic and propionic acid solutions at 1%, 2% and 3% (w/v) for 1 min) and natamycin (50, 100, 150 and 200 mg/L for 1 min) have been seen as effective in suppressing fungal growth on harvested chestnuts.[13] Natamycin is a legal food additive (E235) but, being an antibiotic, could cause allergy and antibiotic resistance in consumers.

Ozone in water is another strong oxidizer. It is a water-soluble gas and, depending on the concentration, it is highly corrosive to equipment including rubber, some plastics, and fibreglass. Its commercial use in contact with the product is differently regulated depending on the country, but generally its use is allowed for storage room or ambient sanitization. The great advantage (and disadvantage) of ozone is its rapid decomposition especially in water (half-life = 15–20 min in clean tap water, but less than 1 min in water containing suspended soil particles and organic matter). This is the case for chestnut dipping water and for this reason water should be filtered frequently. At tap water pH, it does not decompose but it is more effective at low temperature, thus water should be cooled to increase its half-life. Its rapid decomposition does not leave any residue on the product or in the environment. If ozonated water is used as a pre-storage treatment, storage should be also done with the use of ozone gas in a cold room. An ozone/water solution of various concentrations (1.0, 0.5 mg/mL, and 0.25 mg/mL) provided good results in the storage of Chinese chestnuts in plastic bags stored for 180 days in a cold room.[15]

Another effective pre-storage or pre-shipping treatment is carbon dioxide at high concentration (shock treatment). In fact, storage conditions characterized by 80%–100% CO_2 for 3–10 days (depending on outside temperature: the higher it is, the shorter is the time), with an RH of 90%–100% and air circulation of 0.5 kg air/h/kg of chestnuts, controls internal decay and sprout development. No water contact, no drying need, and low electricity consumption represent the great advantages of this method. On the other hand, airtight containers or room and operator precautions are factors to be taken into account.

9.8.3 Storage

The use of low temperature is the most widespread technique of chestnut storage because it is easy to perform and not too expensive, but different studies have shown that the implementation of controlled atmosphere at low temperature can increase storage efficacy.

Below, some general storage conditions are recommended (a detailed list of suggestions and troubleshooting for chestnut storage in cold room can be found in Mencarelli).[8]

- Temperature: 0°C or −1°C (this condition increases sweetness due to starch breakdown);
- RH: 90%–95%;
- Air flow: 0.1–0.2 m/s through chestnuts in bins;
- Cold room volume recirculation: 4 times/h;
- Perforated plastic bins are better and longer lasting, while wooden bins are not suitable (Figure 9.15).

Carbon dioxide has been proposed for cold storage of chestnuts. Chestnuts are not sensitive to high carbon dioxide and it is known to have a fungicidal effect. An atmosphere with low oxygen (1%–3%) and high CO_2 (5%–30%) at ±1°C is considered efficient to keep the high quality of chestnuts in long-term storage.[16,17] Moreover, these storage conditions are able to control rot development in chestnuts with a high rate of mechanical injuries or torch breakage. Nevertheless, this technique is not frequently used due to the costs of the procedure and the management of an airtight cold room (more expensive than regular cold storage).

The use of ozone in a cold room is more successful, mainly due to economic reasons. Ozone does not need a special cold room; the cost of the generator is

FIGURE 9.15   Wooden bins for chestnut storage are not recommended as they are not easily cleaned. (Courtesy of Mastrogregori Aldo, s.r.l.)

low and the maintenance cost is also low. The gas concentration to use is below 1 ppm and can be managed automatically by using a higher concentration during the night and lower during the day in order to reduce the risk for the operators. Today, a very sensitive alarm system exists which advises of potential leakage. In Technical Sheet 9.3 strengths and weaknesses of ozone and controlled atmosphere implementation in chestnut storage are reported.

TECHNICAL SHEET 9.3   STRENGTHS AND WEAKNESSES OF LOW TEMPERATURE COMBINED WITH OZONE CONTROLLED ATMOSPHERE

Physical Conditions of Storage	Strengths	Weaknesses
Low temperature ($\pm 1°C$) and controlled atmosphere (5%–30% CO_2, 1%–3% O_2)	Necessity of an airtight cold room. CO_2 has a bactericide effect on fungi. Lower rot and insect development. Prevention of water loss. Mould reduction. Good taste. Control of rots developing in chestnuts with high rate of mechanical injuries or torch break.	Difficulties in building an airtight cold room. More expensive than the regular cold storage. Necessity to control atmosphere plant in addition to cooling plant.

(Continued)

Physical Conditions of Storage	Strengths	Weaknesses
Use of ozone in cold room	Special cold room not necessary. Low cost of generator. Low costs of maintenance. High sanitizing power for the cold room and for the chestnut, superficially and inside. The longer the storage, the greater the efficiency. Low cost of purchase and management. No residue left in chestnut.	Special care must be taken by the operator. Ozone can oxidize the chestnut pericarp giving a reddish colour. Plastic material and iron can be destroyed.

REFERENCES

1. Foukaraki, S.G., Cools, K., Chope, G.A. and Terry, L.A. 2016. Impact of ethylene and 1-MCP on sprouting and sugar accumulation in stored potatoes. *Postharvest Biology and Technology* 114: 95–103.

2. Botondi, R., Vailati, M., Bellincontro, A., Massantini, R., Forniti, R. and Mencarelli, F. 2009. Technological parameters of water curing affect postharvest physiology and storage of marrons (*Castanea sativa* Mill., Marrone fiorentino). *Postharvest Biology and Technology* 51(1): 97–103.

3. Tweddell, R., Boulanger, R. and Arul, J. 2003. Effect of chlorine atmospheres on sprouting and development of dry rot, soft rot and silver scurf on potato tubers. *Postharvest Biology and Technology* 28(3): 445–454.

4. Vannini, A., Vettraino, A., Martignoni, D., Morales-Rodriguez, C., Contarini, M., Caccia, R., Paparatti, B. and Speranza, S. 2017. Does *Gnomoniopsis castanea* contribute to the natural biological control of chestnut gall wasp? *Fungal Biology* 121(1): 44–52.

5. Donis-González, I.R., Guyer, D.E., Pease, A. and Fulbright, D.W. 2012. Relation of computerized tomography Hounsfield unit measurements and internal components of fresh chestnuts (*Castanea* spp.). *Postharvest Biology and Technology* 64(1): 74–82.

6. Demiate, I.M., Konkel, F.E. and Pedroso, R.A. 2001. Quality evaluation of commercial samples of doce de leite-chemical composition. *Food Science and Technology* 21(1): 108–114.

7. Li, Z. and Thomas, C. 2014. Quantitative evaluation of mechanical damage to fresh fruits. *Trends in Food Science & Technology* 35(2): 138–150.

8. Mencarelli, F. 2001. Postharvest handling and storage of chestnuts. Working document of the project: TCP/CPR/8925, Integrated Pest Management and Storage of Chestnuts in XinXian County, Henan Province, China.

9. Moscetti, R., Haff, R.P., Saranwong, S., Monarca, D., Cecchini, M. and Massantini, R. 2014. Nondestructive detection of insect infested chestnuts based on NIR spectroscopy. *Postharvest Biology and Technology* 87: 88–94.

10. Donis-González, I.R., Guyer, D.E., Fulbright, D.W. and Pease, A. 2014. Postharvest noninvasive assessment of fresh chestnut (*Castanea* spp.) internal decay using computer tomography images. *Postharvest Biology and Technology* 94: 14–25.

11. Jermini, M., Conedera, M., Sieber, T.N., Sassella, A., Schärer, H., Jelmini, G. and Höhn, E. 2006. Influence of fruit treatments on perishability during cold storage of sweet chestnuts. *Journal of the Science of Food and Agriculture* 86(6): 877–885.

12. Kwon, J.-H., Kwon, Y.-J., Byun, M.-W. and Kim, K.-S. 2004. Competitiveness of gamma irradiation with fumigation for chestnuts associated with quarantine and quality security. *Radiation Physics and Chemistry* 71(1–2): 43–46.

13. Panagou, E., Vekiari, S., Sourris, P. and Mallidis, C. 2005. Efficacy of hot water, hypochlorite, organic acids and natamycin in the control of post-harvest fungal infection of chestnuts. *The Journal of Horticultural Science and Biotechnology* 80(1): 61–64.

14. Donis-González, I., Fulbright, D., Ryser, E. and Guyer, D. 2009. Efficacy of postharvest treatments for reduction of molds and decay in fresh Michigan chestnuts. *Acta Horticulturae* 866: 563–570.

15. Zhang, B., Zhu, X. and Lai, J. 2011. Effect of ozone water treatment on preservation of chinese chestnut fruits during storage. *Food Science* 32(16): 361–364.

16. Anelli, G., Mencarelli, F., Nardin, C. and Stingo, C. 1982. conservazione castagne mediante l'impiego delle atmosfere controllate. *Industrie alimentari*.

17. Rouves, M. and Prunet, J. 2002. New technique for chestnut storage: Effects of controlled atmosphere. *Infos Ctifl* 186: 33–35.

10 Coppice Woodlands and Chestnut Wood Technology

Maria Chiara Manetti, Enrico Marcolin,
Mario Pividori, Roberto Zanuttini,
and Marco Conedera

CONTENTS

10.1 HISTORY OF CHESTNUT COPPICE MANAGEMENT

The tradition of managing chestnut forests as coppice for timber production has been developed almost exclusively in relation to the European sweet chestnut and its high resprouting capacity[1] (Figure 10.1). First written evidences of coppice management for chestnut timber production date back to the *Historia plantarum* (Inquiry to plants, H.P.) by the ancient Greek writer Theophrastus (~370 B.C.–287 B.C.), whereas unambiguous reports on chestnut coppice stands are provided in the *De re rustica* (On farming) and *Naturalis Historia* (Natural History) by the post-Christian Latin writers Columella (first century) and Pliny the Elder (23/24 A.D.–79 A.D.), respectively. The biological and economical sustainability of the chestnut coppice probably represents the main reason why Romans developed the idea of cultivating it.[1]

In the Middle Ages coppice represented the dominant forest management approach in a large portion of central Europe.[2] Evidence of an industrial approach to the chestnut coppicing exists starting in the eighteenth century for southeast England

FIGURE 10.1 Abundant stool resprouts in a young simple chestnut coppice in southern England. (Courtesy of Conedera, M.)

(i.e., Kent, Surrey and Sussex), where chestnut coppice poles were produced as support for the hopbines in the hop industry, as pit props for mines, and later, for wired paled fences[3] (Figure 10.2).

Most traditional chestnut coppices shared (and some still share) the common features originally deriving from the Roman approach of being managed in relatively short rotations and without any silvicultural treatment (e.g., thinning), in order to produce small to medium-sized poles, fire wood, and charcoal.[1] This type of management experienced a deep crisis with the changes of the socioeconomic

FIGURE 10.2 Wired chestnut wood fences. (Courtesy of Conedera, M.)

structures in rural areas and the progressive vanishing of the request for chestnut poles on the market.[4] In the 1950s the international community reacted by creating a group of chestnut experts in charge of analyzing the structural and current problems (i.e., occurrence of new pathogens such as the chestnut blight) of the European chestnut cultivation. In the following decades, new silvicultural concepts related to the chestnut coppice cultivation for the production of high quality wood have been proposed going from the extension of the rotation period, to the application of early, intense, and frequent thinning intervention, to the single tree-oriented silviculture.[5,6]

At present the European chestnut-growing area devoted to timber production is 1.78 million hectares. Management systems and intensities highly vary among countries although most of the area (79%) is managed as coppice stands.[4]

10.2 MAIN CHESTNUT COPPICE MANAGEMENT SYSTEMS

Three main management options can be defined in relation to chestnut coppice systems.

Simple coppice – All trees and shoots of the stand are cut (Figure 10.1). From the ecological point of view this is the most suitable system for a light-demanding species like the chestnut tree. Chestnut resprouts take high advantage from the light provided by the complete clear cutting and removal of the mature generation. Intense and multi-generational simple coppice stands usually reach a density of more than 400 stools ha^{-1} that ensures a good covering of the soil.[7]

Coppice with standards – Some individuals (standards) are left to grow longer than one coppice cycle (Figure 10.3) in order to assure a permanent minimal soil cover, to gain timber of larger dimensions, and to provide seed dispersal. Seed production by standards may be important not only for the natural regeneration, but also as fodder source for wild and domestic animals.[8]

Pollarding – Radical pruning (in some cases practiced even as topping, Figure 10.4). Not a very common approach and locally restricted in relation to historical traditions (i.e., Spain, southern Switzerland). It is usually connected to specific purposes such as combining fruit and timber production, allowing intercropping, phytosanitary cuttings, conservation of ownership, etc.

10.3 STRENGTHS AND WEAKNESSES OF CHESTNUT COPPICES

A detailed knowledge of the strengths and weaknesses of the coppice system is a prerequisite for evaluating its potential in terms of multiple ecosystem services such as protective function, environmental and biodiversity conservation, quality wood production, and maintenance of the social component and the cultural heritage.

FIGURE 10.3   Coppice with standards in central Italy. (Courtesy of Manetti, M.C.)

FIGURE 10.4   Pollarding on old chestnut trees in southern Switzerland. (Courtesy of Conedera, M.)

Two main categories of factors may be considered in a SWOT analysis:

1. *Intrinsic factors*, which are determined by the environment and can hardly be influenced by forestry approaches or changes in the silvicultural management;
2. *Induced factors*, which may be impacted by targeted social, political, or silvicultural approaches, enhancing or diminishing the value of a chestnut coppice system.

10.3.1 POSITIVE INTRINSIC FACTORS

- *Growth rate* – As a general rule, chestnut coppices display remarkable growth rates that may easily surpass 10 m^3 ha^{-1} in annual current increment. This intrinsic strength can be further enhanced by programming timely and regular silvicultural treatments, in order to keep constant growth rates over time.[5]
- *Resprouting capacity* – The resprouting capacity of the chestnut tree is very high and almost inexhaustible (Figure 10.1). Its resprouting ability is high also in advanced age and seems to be insensitive to the season of the silvicultural treatment.[9]
- *Root system physiology* – The stools renews the root system after each coppicing.[10]
- *Sexual maturity* – Under optimal environmental conditions (good site, sufficient precipitation, light-exposed upper crown), chestnut shoots show an early seed production (<10 years). Seeds usually display a high germination rate.
- *Regeneration* – Advanced and post-harvesting regeneration is abundant under suitable environmental conditions[11,12] (Figure 10.5).

FIGURE 10.5 Post-thinning natural regeneration in a 20-year-old chestnut coppice. (Courtesy of Conedera, M.)

- *Wood characteristics* – The wood is of good technological and aesthetic quality and is thus suitable for many uses (see specific chapter below). In particular it displays a natural resistance to decaying agents thanks to the high tannin content that makes it particularly useful for outdoor applications.

10.3.2 POSITIVE INDUCED FACTORS

- *Availability* – Chestnut stands devoted to wood production are broadly distributed across the European chestnut countries,[4] with some hotspots in France, Italy, and Spain (Table 10.1).
- *Silvicultural plasticity* – The long-lasting, high growth rate of the chestnut tree allows adaptation of the final rotation time to specific production needs, so as to diversify the produced timber assortments or to anticipate the profits in case of thinning activities.[6,13]
- *Cultural and social context* – In most European chestnut countries, the chestnut wood is part of the cultural heritage. This represents a huge

TABLE 10.1
Chestnut Area Devoted to Wood Production in Europe

Country	Chestnut Area (ha)		
	Total Chestnut	Wood Production	%
France	1,020,500	920,500	90
Italy	765,837	497,870	65
Spain	137,627	99,948	73
Georgia	48,000	48,000	100
Russian federation	40,000	40,000	100
Portugal	53,509	33,900	63
Greece	33,651	33,051	98
Slovenia	30,185	30,000	99
Switzerland	27,100	23,700	87
United Kingdom	18,788	18,788	100
Croatia	15,000	15,000	100
Germany	4,400	4,400	100
Turkey	28,892	3,614	13
Bosnia-Herzegovina	3,057	3,057	100
Romania	2,990	2,890	97
Bulgaria	2,960	2,100	71
Slovakia	1,505	1,318	88
Hungary	2,000	1,100	55
Belgium	1,050	1,050	100
Netherlands	50	50	100
TOTAL (ha)	**2,237,101**	**1,780,336**	**80%**

Source: Conedera M. et al., *Ecol. Mediterr.*, 30, 179–193, 2004b.

opportunity to develop local production chains in rural and marginal territories based on social and economic sustainability.[14,15]

- *Multifunctionality* – The direct and indirect ecosystem services provided by chestnut coppices go far beyond wood production. Besides the secondary non-woody products such as fruits, honey, and mushrooms, important general services are represented by the protection against natural hazards (slides, rock fall), water cycle regulation, carbon sinking, and biodiversity.[16,17]

10.3.3 NEGATIVE INTRINSIC FACTORS

- *Ring shake* – Chestnut wood presents a predisposition of cracking along the ring (Figure 10.6) when internal growth-induced tensions are released through the tree falling or the subsequent drying process.[18,19] This intrinsic wood characteristic represents a major problem for producing quality wood in chestnut coppices that can be only partially mitigated by appropriate silvicultural management (i.e., timely and regular thinning that guarantees regular and sustained growth).
- *Drought sensitivity* – Despite its general broad environmental plasticity, the chestnut tree is highly sensitive to drought periods, especially during the growing season.[20] The drought-induced low and irregular growth enhances the ring shake risk, whereas prolonged or repeated drought episodes highly impact the overall wood production and, in extreme cases, the tree health conditions.
- *Pests and pathogens* – Unlike the Asian chestnut tree species that co-evolved with most of the pathogens affecting the genus *Castanea*, the European chestnut is quite sensitive to diseases such as chestnut blight, ink disease, and partially, leaf spot. Most of the diseases are not lethal for

FIGURE 10.6 Chestnut wood section heavily affected by ring shake. (Courtesy of WSL Cadenazzo.)

the chestnut tree (although ink disease may represent an exception in this respect), but may heavily reduce the growth performance and the quality of the stands. This is also the case for the major chestnut tree pest, the Asian gall wasp, which limits the normal vegetation of the trees, severely affecting crown architecture and growth performance.

10.3.4 NEGATIVE INDUCED FACTORS

*   *Stool biology* – Despite the elevated resprouting capacity, overaged stools tend to lose their vitality over time. When abandoned to a natural evolution far beyond the coppice rotation time, overaged stools tend to develop a disproportion between the continuously increasing aerial part and the root system, which may cause progressive instability and eventually the uprooting of the concerned stools[21] (Figure 10.7).
*   *Competition and invasibility* – Chestnut stools and trees in unmanaged coppices are not very competitive and tend thus to be very prone to the colonization by spontaneous forest species, including some alien species[22,23] (Figure 10.8). Chestnut coppice thus requires a regular and targeted management plan to be preserved.
*   *Socioeconomic structural weaknesses* – Areas potentially devoted to chestnut coppicing often display structural problems that partially compromise the effectiveness of the forest management.[15] These problems include fragmented private ownerships, lack of coordination and corporate association among operators, high processing and personnel costs with respect to the

FIGURE 10.7   Uprooted stool in an over-aged coppice stand in southern Switzerland. (Courtesy of Bomio, P.)

FIGURE 10.8   Freshly thinned chestnut coppice invaded by the alien neophitic tree *Paulownia tomentosa* in southern Switzerland. (Courtesy of Conedera, M.)

produced value, lack of clear silvicultural objectives and management strategies, and lack of information among professionals about positive characteristics and opportunities related to the coppice management.

- *Fire risk* – Chestnut coppices are rich in flammable necromass that may represent a very dangerous fuel in case of forest fires, especially during the vegetation rest. In fire-prone areas this may be an additional risk factor for the wood production in general and of quality timber in particular.[24]

10.4   NON-WOOD PRODUCTS AND ECOSYSTEM SERVICES

- *Non-wood products* – Important non-wood products of chestnut coppices are mainly represented by chestnut honey production and the possibility of collecting mushrooms.[25,26]
- *Protective function* – Dense chestnut coppice stands assure a good interception of the precipitation and a stabilization of the soil through root reinforcement. The high number of stems provides also a very good protection against rolling stones. All these effects, however, shortly vanish in the early years after coppicing, especially for simple coppices.[10]

- *Ecological functions and biodiversity* – Naturalness and biodiversity in chestnut coppices highly depend on the scale considered. At landscape scale, very mobile fauna-like birds benefit from the structure diversification due to the different stand development stages. At stand scale, habitat diversification highly depends on the specific system (coppice with standards or not), rotation time (size of the trees), and intensity of treatments (dead wood resulting from the interventions).[27]
- *Carbon sequestration* – According to the rotation time and the silvicultural approach adopted, chestnut coppices may significantly contribute to carbon fixation.[28,29]
- *Recreational and didactic activities* – The coppice system is based on vegetative resprouts and therefore constitutes a fascinating topic for people interested in botany. Short-rotation coppices are not easily accessible, but their presence may contribute to enhancing the visibility of the landscape from existing trails, whereas mature coppices have a very high recreational value that may be compared with that of high forests.

10.5   WOOD AND RELATED PRODUCTS

10.5.1   WOOD PROPERTIES

Chestnut wood has interesting technological properties that make it easy to process and suitable for several uses where aesthetic or performance requirements prevail.[30,31] The most appreciated properties (Technical Sheet 10.1) are the good natural durability of its heartwood, the high mechanical resistance in relation to density,[32] and the low shrinkage. It additionally displays a light and generally uniform color (with shades from golden yellow to yellowish-brown) and an evident grain that give an appreciated surface appearance.

- *Macroscopic features* – The wood displays a gray-brown and homogeneous color throughout the cross-section. It clearly differentiates with age in a thin sapwood (formed by a few peripheral rings) of light color and a brown heartwood, sometimes with dark and irregular streaks. The growth rings, generally wider than 3 mm, are well marked and perceptible with the naked eye due to the presence of the characteristic large early-wood vessels in the initial ring-porous portion of the ring (Figure 10.9). Compared to oak, chestnut wood is matte and easily distinguishable because of the lack of visible rays (flakes effect) on longitudinal surfaces.
- *Workability* – The major critical characteristic of chestnut is the ring shake, a crack developing along a longitudinal-tangential plane between two adjacent growth rings that generally compromises the possibility of further processing (Figure 10.6). The presence of ring shake seems to depend mainly on trauma or growing stresses and can be influenced by silvicultural practices.[13] Furthermore, chestnut stems frequently show defects of shape and several knots.

TECHNICAL SHEET 10.1   WOOD PROPERTIES
Main wood properties (determined on small size and defect-free specimens)

Property	Description
Density (in green condition)	From 700 to 1100 kg/m³
Density (at 12% wood moisture content)	From 470 to 700 kg/m³ – average value: 580 kg/m³
Acidity	High (pH 3.6)
Total linear shrinkage	Axial: 0.6%, radial: 4.1%, tangential: 6.1%, average volumetric shrinkage: 10.8%
T/R anisotropy ratio	Around 1.5
Compression strength	From 21 to 64 N/mm², average value: 51 N/mm²
Bending strength	From 50 to 140 N/mm², average value: 86 N/mm²
Modulus of elasticity on bending	From 8450 to 14400 N/mm², average value: 11380 N/mm²
Shear strength	From 5.7 to 9.2 N/mm², average value: 7.3 N/mm²
Shock resistance and hardness	From low to medium
Durability to basidiomycetes	Class 2 (EN 350)
Durability to xylophagous insects	Class D (EN 350)
Durability to termites	Class M (EN 350)
Treatability	Class 4 – extremely difficult
Use class complying with natural durability	Class 3 – for external use (but not in ground contact or with permanent humidification)

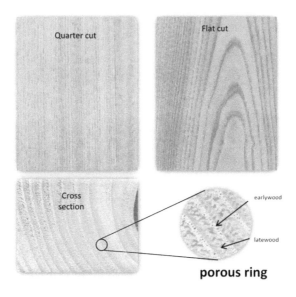

FIGURE 10.9   Appearance of chestnut wood in the main anatomical sections. (Courtesy of Zanuttini, R.)

The natural seasoning is rather slow but without particular drawbacks; artificial drying requires temperatures not too high and is often difficult, especially for thicker sawn timber. Planning and sanding are not always easy. Due to the considerable porosity of the surface, painting requires a generous amount of primer (filler) but provides good results and an excellent finish. Bonding does not show particular difficulties, but the use of alkaline adhesives or fillers should be avoided since they cause stains on the semi-finished products. Joints with nails or screws are easily achievable but, due to the high flexibility of the wood, they are mostly poorly sealed. Due to its acidity, chestnut wood tends to corrode metals in case of high humidity. The reaction of the tannins determines the formation of black-bluish spots on the wooden surfaces that come in contact with ferrous materials. The natural durability is poor in the sapwood, which is attacked by fungi and insects, but very good in the heartwood, what makes it suitable for external uses. The impregnation with preservatives is not easy; however, it is considered superfluous for the heartwood.

10.5.2   MAIN PRODUCTS AND USES

- *Pole assortments* – The poles for vineyards and orchards still represent an important outlet for chestnut wood: the good mechanical and durability properties guarantee a long period of service even in critical climatic exposure (Figure 10.10). Depending on local habits, chestnut poles are sold debarked or not, with or without tip, by weight, length, or number of pieces.

FIGURE 10.10   Example of chestnut poles for agricultural uses. (Courtesy of Segheria Valle Sacra.)

FIGURE 10.11   The natural durability of chestnut wood makes it suitable for bioengineering applications. (Courtesy of Segheria Valle Sacra.)

The possibility of obtaining a regular shape of the poles through the use of suitable machinery opens new prospects for higher value-added uses such as in urban furniture, garden sector (for fences, benches, etc.), and for the market of environmentally friendly products in general. Chestnut poles are used also in bioengineering works, for instance in the consolidation of slopes subject to erosion (Figure 10.11) and as guardians in arboriculture plants, whereas split small poles are tied together to produce the typical wired fences that are very popular in England and other northern European countries (Figure 10.2).

- *Beams* – Beams on solid wood for various carpentry uses can be divided into two categories: timber with a constant rectangular cross section along the entire length (with a limited tolerance of localized wane) and 'Uso Fiume', realized by mechanically squaring on all four sides of the trunk in order to obtain a constant section with continuous and wider wanes (Figure 10.12). According to the EU Construction Products Regulation n. 305/2011 (CPR), the solid wood assortments for structural use (beams, boards, and match-boards) are subject to the obligation of strength grading (visual or machine) and CE marking.[33,34]
- *Sawn wood* – Refers to lumbers destined for the production of furniture, joinery elements, doors, and windows (Figure 10.13), external cladding (sometimes still in form of split shingles and staves for craft or industrial construction of vats, barrels, and barriques for preserving, refining, or fla-voring wines, liqueurs, and vinegars). The nice appearance and the light color shade of the wood make the chestnut attractive for solid or prefin-ished wood floors (with an upper layer of thin sawn wood), even if its hard-ness is lower with respect to oak wood. Recently, modified wood products made of solid wood (boards and planks) heat-treated at high temperatures

FIGURE 10.12 Chestnut beams 'Uso Fiume' for structural use in traditional carpentry. (Courtesy of Segheria Puppo.)

FIGURE 10.13   Rustic door on chestnut wood. (Courtesy of Zanuttini, R.)

(from 160°C to 210°C) in order to improve some wood properties and make it suitable for uses as external coatings and floors have raised a wide interest.

- *Sliced veneers* – Chestnut wood provides decorative sliced veneers for various furnishing components as an alternative to oak. Rotary cutting, however, is generally not performed.
- *Wood-based panels and glulam* – Solid wood panels (made up of one or three layers, with entire or jointed strips) are mainly addressed to the furnishing industry (rustic components such as table tops, windows, or doors). Panels and glulam profiles are often used to replace uprights and morals of solid wood. Other products of joists or cross laminated timber are used as structural elements for the renovation of rural buildings or as sound-absorbing barriers. Secondary assortments issued from chestnut branchwood and processing waste are used blended with other hardwoods for the production of medium-density fiberboard (MDF), whereas the production of oriented strand boards (OSB) chestnut panels is under testing.
- *Tannins* – The tannins content of chestnut wood (including the bark) varies between 8% and 12% and is usually obtained from low quality timber with a diameter of at least 10 cm. Tannins are used in the adhesive and chemical sectors and as food supplements for animals. Chemical extracts can be used to tan hides and skins, to improve the resistance of particle boards, to refine wines (barrels), and for many other dieting and cosmetic purposes.[35] The wooden waste material after tannins extraction is used for the production of pellets.
- *Assortments for energy* – The use of chestnut wood for fuel in the form of firewood or wood chips is not optimal since the presence of tannin makes the combustion difficult (it burns with cracking and frequent projection of embers). However, this use is common where the resource is abundant or where other more suitable wood species are not available. Chestnut timber for chips production is often sold at very low prices to feed district heating centers or biomass cogeneration plants.

The main products are summarized in Table 10.2.

TABLE 10.2

Main Wood Products of Chestnut Coppices

Assortments	Diameter (cm)	Length (m)	Use
Beams	>20	>4	Carpentry, building
	>20	2–4	Joinery, furniture
Joists	17–24	4–5	Carpentry, building
	12–16	2–4	
Bioengineering	15–25	4	Environmental restoration
Poles	6–15	Variable	For agriculture, support,
	2–5		and fences

10.6   PRESENT AND FUTURE SILVICULTURAL TRENDS

At present, three different management options for chestnut coppices have been detected.

1. Traditional management (business as usual) with short/medium rotation and without silvicultural treatments such as thinning. Management objectives are clear and consolidated, but often the final product does not allow an economically sustainable management. The risk is high for such coppices to get with time over-aged and then abandoned.
2. The non-management option consists of the abandonment of an active silviculture. In this situation, only sporadic harvesting activities exist on mature stools. Such aged and abandoned coppices are frequently subjected to recrudescence of chronic chestnut diseases and loss of stump vitality with progressive replacement of the chestnut tree by native or even alien tree species.
3. Coppices managed for quality wood production. The traditional coppice approaches based on short rotations and no thinning activities only aim at producing poles. They fail, however, to exploit the whole potential of the chestnut tree. Improvements of the chestnut coppice silviculture should aim at producing bigger log dimensions (>30 cm) by extending the rotation time and planning timely and frequent thinning interventions that allow the trees to grow in a strong and regular way.

   In order to switch from traditional to quality wood coppice management (from option 1 to option 3) or to restore abandoned chestnut coppices (from option 2 to option 3), the following prerequisites are needed:
   Site characteristics – Quality wood silviculture should be limited to good sites (i.e., continuous water supply, no drought stress in summer). Site index or dominant height are good parameters to assess site productivity.
   Stand characteristics – Species composition (i.e., chestnut proportion) and density (number of stools and crown cover) are important characteristics to evaluate the potential of present chestnut coppice generation. Ideal stand density may vary as function of the site fertility but usually ranges from 800 to 1,200 stools per hectare at the age of first thinning.
   Accessibility and forestry organization – Coppice stands should be easily accessible for wood hauling, ownership should be suitable (no excessive fragmentation), and forestry companies should be credited to do the work properly. Finally, the local and regional timber transformation

TABLE 10.3
Silvicultural Options Suggested for Chestnut Coppices Related to Productivity and Timber Assortments

Productivity	Rotation Time (years)	Thinning (nr.)	Silvicultural Approach	Mean Diameter (cm)	Timber Assortments
Low	Ageing or conversion into a mixed forest with valuable and suitable tree species				
Medium–Low	16–25	0–1	Stand silviculture	12–15	Poles
Good–Medium	25–35	2–3		>20	Poles, beams
High–Good	>40	≥3		>30	Poles, beams, sawlogs
High–Good	30–50	≥3	Tree-oriented silviculture	>35	Poles, beams, sawlogs

companies and an effective marketing should ensure a real market demand and product processing.

Additional frame conditions – Occurrence of ring shake in the area, as well as pathogen (ink disease and chestnut blight in particular) and pest (gall wasp) pressures represent secondary, but possibly very important, additional frame conditions that should be considered.

As reported in Table 10.3, selecting target products as a function of site fertility automatically implies to define a corresponding rotation time and highly restricts the actionable silvicultural options that are basically:

Frequency and intensity of thinning – Selective thinning aims at providing enough space, water, and nutrients to the target stems (candidates), which usually respond with enhanced growth increments and a better quality in the shape of the stems.

Thinning treatment – Thinning may be applied on the whole stand in a homogeneous way (stand silviculture, Figure 10.14) or may concentrate on the best candidates, i.e., dominant individuals of the upper layer displaying outstanding characteristics in term of stem quality, dimension, vigor, and spatial distribution (optimal distance between candidates). Such candidates are recognized at an early stage and consequently favored through targeted thinning (tree-oriented silviculture, Figure 10.15). Technical Sheet 10.2 summarizes the main characteristics of the two approaches.

FIGURE 10.14 Coppice subjected to thinning treatments based on stand silviculture approach. (Courtesy of Conedera, M.)

FIGURE 10.15 Coppice subjected to tree-oriented silviculture. (Courtesy of Conedera, M.)

TECHNICAL SHEET 10.2   SILVICULTURAL APPROACHES

Main silvicultural approaches for food quality wood production in coppices.

Silvicultural Aspects	Stand Silviculture	Tree-oriented Silviculture
First intervention	After 7–15 years, according to the site fertility, but in any case before the competition among shoots starts	After 8–10 years, according to the site fertility, but in any case when the differentiation among shoots starts to be visible
Number of candidates	The number of target trees is not defined. The 20%–30% of basal area is removed at every thinning	80–100 ha^{-1} in Italy and Switzerland Ca. 80 ha^{-1} in France
Thinning type	Selective thinning aiming at eliminating the most competitive trees with respect to the dominant ones	Making the crown of the candidate completely free (*déturage*) by eliminating all the main competitors
Thinning frequency	Every 4–7 years according to site fertility, but before the canopy closure limits the stem increment of candidates	According to the needs, but 2–3 times between 10 and 25 years on average
Pruning	None	Green pruning of the candidates in the second year after the silvicultural interventions. Epicormic shoots should be very limited and dying back after few year.
Additional measures	None	Single tree protection against ungulates for the candidates

Source: Manetti, M.C. et al., *Adv. Hortic. Sci.*, 20, 65–69, 2006; Lemaire, J. et al., *Foret Entreprise*, 179, 68 p, 2008.

REFERENCES

1. Conedera, M., Krebs, P., Tinner, W., Pradella, M. and Torriani, D. 2004a. The cultivation of *Castanea sativa* (Mill.) in Europe, from its origin to its diffusion on a continental scale. *Vegetation History and Archaeobotany* 13(3): 161–179.
2. Squatriti, P. 2013. *Landscape and Change in Early Medieval Italy: Chestnuts, Economy, and Culture*. Cambridge, UK: Cambridge University Press.
3. Bartlett, D.M.F. 2011. The history of coppicing in South East England in the modern period with special reference to the chestnut industry of Kent and Sussex. PhD thesis, University of Greenwich.
4. Conedera, M., Manetti, M., Giudici, F. and Amorini, E. 2004b. Distribution and economic potential of the Sweet chestnut (*Castanea sativa* Mill.) in Europe. *Ecologia Mediterranea* 30(2): 179–193.
5. Bourgeois, C., Sevrin, E. and Lemaire, J. 2004. *Le châtaigner, un arbre, un bois*. 2nd revised ed., Paris, France: Institut pour le développement forestier.

6. Manetti, M.C., Amorini, E. and Becagli, C. 2006. New silvicultural models to improve functionality of chestnut stands. *Advances in Horticultural Science* 20(1): 65–69.
7. Gallardo-Lancho, J.F. 2001. Distribution of chestnut (*Castanea sativa* Mill.) forests in Spain: Possible ecological criteria for quality and management (focusing on timber coppices). *Forest Snow and Landscape Research* 76(3): 477–481.
8. De Vasconcelos, M.C., Bennett, R.N., Rosa, E.A. and Ferreira-Cardoso, J.V. 2010. Composition of European chestnut (*Castanea sativa* Mill.) and association with health effects: Fresh and processed products. *Journal of the Science of Food and Agriculture* 90(10): 1578–1589.
9. Giudici, F. and Zingg, A. 2005. Sprouting ability and mortality of chestnut (*Castanea sativa* Mill.) after coppicing. A case study. *Annals of Forest Science* 62: 513–523.
10. Dazio, E., Conedera, M. and Schwarz, M. 2018. Impact of different chestnut coppice managements on root reinforcement and shallow landslide susceptibility. *Forest Ecology and Management* 417: 63–76.
11. Zlatanov, T., Velichkov, I., Georgieva, M., Hinkov, G., Zlatanova, M., Gogusev, G. and Eastaugh, C. 2015. Does management improve the state of chestnut (*Castanea sativa* Mill.) on Belasitsa Mountain, southwest Bulgaria? *iForest-Biogeosciences and Forestry* 8(6): 860–865.
12. Manetti, M.C., Becagli, C., Pelleri, F., Pezzatti, G.B., Pividori, M., Conedera, M. and Marcolin, E. 2018. Assessing seed regeneration in chestnut coppices: A methodological approach. *Annals of Silvicultural Research* 42(2): 85–94.
13. Lemaire, J., Lempire, R., Cousseau, G. and Pichard, G. 2008. Sylviculture du châtaigner. *Foret Entreprise (numéro spécial)* 179: p. 68.
14. Becagli, C., Amorini, E., Fratini, R., Manetti, M.C. and Marone, E. 2010. Problems and prospects of the chestnut timber chain in Tuscany. *Acta Horticulturae* 866: 693–702.
15. Willis, S. and Campbell, H. 2004. The chestnut economy: The Praxis of neo-peasantry in rural France. *Sociologia Ruralis* 44(3): 317–331.
16. Covone, F. and Gratani, L. 2006. Age-related physiological and structural traits of chestnut coppices at the Castelli Romani Park (Italy). *Annals of Forest Science* 63: 239–247.
17. Patricio, M.S., Nunes, L.F. and Pereira, E.L. 2014. Evaluation of soil organic carbon storage in a sustainable forest chestnut management context. *Acta Horticulturae* 1043(161–165).
18. Fonti, P., Bräker, O.U. and Giudici, F. 2002. Relationship between ring shake incidence and earlywood vessel characteristics in chestnut wood. *IAWA Journal* 23(3): 287–298.
19. Mutabaruka, C., Woodgate, G.R. and Bucley, G.P. 2005. External and internal growth parameters as potential indicators of shake in sweet chestnut (*Castanea sativa* Mill.). *Forestry* 78: 175–186.
20. Afif-Khouri, E., Alvarez-Alvarez, P., Fernandez-Lopez, M.J., Oliveira-Prendes, J.A. and A., C.-O. 2011. Influence of climate, edaphic factors and tree nutrition on site index of chestnut coppice stands in north-west Spain. *Forestry* 84(4): 385–396.
21. Vogt, J., Fonti, P., Conedera, M. and Schroder, B. 2006. Temporal and spatial dynamics of stool uprooting in abandoned chestnut coppice forests. *Forest Ecology and Management* 235: 88–95.
22. Conedera, M., Stanga, P., Oester, B. and Bachmann, P. 2001. Different post-culture dynamics in abandoned chestnut orchards and coppices. *Forest Snow and Landscape Research* 76: 487–492.
23. Pezzi, G., Maresi, G., Conedera, M. and Ferrari, C. 2011. Woody species composition of chestnut stands in the Northern Apennines: The result of 200 years of changes in land use. *Landscape Ecology* 26(10): 1463–1476.

24. Pezzatti, G.B., Bajocco, S., Torriani, D. and Conedera, M. 2009. Selective burning of forest vegetation in Canton Ticino (southern Switzerland). *Plant Biosystems* 143(3): 609–620.
25. Diamandis, S. and Perlerou, C. 2001. The mycoflora of the chestnut ecosystems in Greece. *Forest Snow and Landscape Research* 76(3): 499–504.
26. Gavaland, A. and Pelletier, S. 2006. Multi-purpose management of chestnut: The French situation. *Advances in Horticultural Science* 20: 70–81.
27. Guitián, J., Munilla, I., Guitián, J., Garrido, J., Penin, L., Dominguez, P. and Guitián, L. 2012. Biodiversity in chestnut woodlots: Management regimen vs woodlot size. *Open Journal of Forestry* 2(4): 200–206.
28. Martínez-Alonso, C. and Berdasco, L. 2015. Carbon footprint of sawn timber products of *Castanea sativa* Mill. in the north of Spain. *Journal of Cleaner Production* 102: 127–135.
29. Prada, M., Bravo, F., Berdasco, L., Canga, E. and Martínez-Alonso, C. 2016. Carbon sequestration for different management alternatives in sweet chestnut coppice in northern Spain. *Journal of Cleaner Production* 135: 1161–1169.
30. Fioravanti, M., Lemaire, J. and Togni, M. 2010. Enhancement of timber production, pp. 44–57. *Chestnut (Castanea sativa): A Multipurpose European Tree*. Brussels 30 September–01 October.
31. Negro, F., Cremonini, C., Zanuttini, R. and Dezzutto, S. 2017. *Il legno di Castagno. Conoscerne il valore, (ri)scoprirne le potenzialità*. Compagnia delle Foreste, Arezzo, Italy.
32. Romagnoli, M., Cavalli, D. and Spina, S. 2014. Wood quality of chestnut: Relationships between ring width, specific gravity and physical and mechanical properties. *Bioresources* 9: 1132–1147.
33. Brunetti, M., Nocetti, M. and Burato, P. 2013. Strength properties of chestnut structural timber with wane. *Advanced Materials Research* (778): 377–384.
34. Zanuttini, R. 2014. *Il legno massiccio. Materiale per un'edilizia sostenibile*. Compagnia delle Foreste, Arezzo, Italy.
35. Aires, A., Carvalho, R. and Saavedra, M.J. 2016. Valorization of solid waste from chestnut industry processing: Extraction and optimization of polyphenols, tannins and ellagitannins and its potential for adhesives, cosmetic and pharmaceutical industry. *Waste Management* 48: 457–464.

11 Diseases

Paolo Gonthier and Cécile Robin

CONTENTS

11.1 INTRODUCTION TO MAJOR AND MINOR DISEASES OF CHESTNUT: HISTORY AND IMPACT

All over the world, chestnut trees are hosts of multiple microorganisms, most of them being fungi and fungi-like organisms. Some of them are beneficial, like mycorrhizal fungi, and others have a negative effect on the host. This effect can be minor thanks to natural selection; indeed, when occurring in the same area, chestnuts and these pathogenic fungi or fungi-like organisms have coevolved to reach a stable equilibrium. By contrast, when these pathogens are introduced in a novel area and encounter chestnut species with which they did not coevolve, they may have a highly detrimental effect on chestnut trees. In Europe and North America, cultivation and conservation of *Castanea* species have been tremendously impacted by two introduced diseases caused by pathogens originating in Asia. These diseases, called ink disease and chestnut blight, primarily affect different organs, i.e. roots or trunk and branches, respectively, but both cause tree decline or death and are thus the most damaging chestnut diseases. Other pathogens, mainly associated with the spoilage of nuts, have important economic impact on nut production.

The ink disease is caused by the fungi-like organisms *Phytophthora cambivora* and *P. cinnamomi*, which both belong to the Oomycota, a large group of microorganisms that look like fungi. Although the first devastating epidemics in Europe were reported as soon as the nineteenth century, these two soil-borne pathogens are thought to have been present in Europe since at least 1726. Outbreaks of *P. cinnamomi* were also reported in the United States between 1825 and 1875. A resurgence of ink disease threatening the cultivation of chestnut in large areas of Europe has been observed since the 2000s. Primary symptoms are root lesions, which can spread to the collar, from which inky fluid exudates often flow, giving the name to the disease. Secondary symptoms are general crown dieback including eventual overall decline, occurring more or less quickly depending on tree resistance and environmental conditions.

The chestnut blight is caused by *Cryphonectria parasitica*, an ascomycete fungus native to eastern Asia and introduced in the early 1900s into North America and then into Europe. The disease devastated the native *C. dentata* forests in the United States, spreading in about 30 years from a single introduction site in New York City throughout the native range of its host. It has been estimated that about 3.5 billion chestnut trees died at that time in United States because of the disease. The symptoms are cortical cankers, resulting in the wilting (blight) of the distal part and in epicormic sprouts which develop below the canker. The fungus has been introduced in Europe from North America and from Asia and is now established in all European countries where *C. sativa* is growing. Despite the high prevalence of the disease in Europe, its severity is lower than in North America, especially in the oldest disease hot spots where most of the cankers heal. This is mainly due to the occurrence in Europe of a virus (Cryphonectria Hypovirus-1, CHV1) which infects *C. parasitica* and reduces its aggressiveness.

Major diseases of chestnut also include nut rots, which are associated with pre-harvest or post-harvest infection by a number of fungi, including *Ciboria batschiana*, causing a black rot, and *Phomopsis* spp., associated with a mummification of nuts.[1] While nut rots have been documented as significantly detrimental for chestnut growers only occasionally or locally, since the mid-2000s, an increased incidence of rotten nuts has been observed in Europe and Australasia which was associated with a previously unknown disease caused by a newly described fungus, the ascomycete *Gnomoniopsis castaneae*.[2] Depending on the location and year, substantial yield losses due to *G. castaneae* occurred, making *G. castaneae* by far the most detrimental nut rot agent of chestnut. For instance, levels of disease incidence as high as 93%, 91%, and 72% were locally reported in northern Italy, Switzerland, and Australasia, respectively.[2]

Chestnut can also be affected by other minor diseases, either root rot diseases or foliar diseases. Such root rots are caused by necrotrophic wood destroying fungi, including the basidiomycetes *Armillaria mellea* and *A. bulbosa*, and the ascomycete *Rosellinia necatrix*. Although these fungi are regarded only as minor chestnut pathogens, they can be locally important as they may result in tree decline and mortality. In general, foliar diseases, including the powdery mildews caused by the biotrophic ascomycetes *Erysiphe* spp., are rarely significantly detrimental to chestnut, except in nurseries. However, leaf spots caused by the necrotrophic ascomycete *Mycosphaerella maculiformis* may be locally damaging in orchards as well.

In the case of chestnut, wood decays have been traditionally regarded as of marginal importance because generally they do not affect significantly the vitality of trees and nut production. However, they may play a prominent role wherever wood is retrieved from chestnut, including forest stands, coppices, and even old orchards.

11.2 SYMPTOMS AND DIAGNOSIS

Regardless of the disease agent and of the organ of the tree colonized by the pathogen, infection may result in the appearance of symptoms in the crown such as leaf yellowing and, in general, symptoms of decline and dieback. This is true not only with pathogens associated with branches and shoots, but also with pathogens associated with roots. Therefore, the above symptoms should be regarded at best as indicative, an accurate diagnosis requiring a thorough analysis of specific symptoms of disease at the organ level often followed by laboratory assays. Specific symptoms of the most common diseases of chestnut, either major or minor, in each organ of the tree as well as guidance for diagnosis are reported below.

11.2.1 ROOTS AND TREE COLLAR

11.2.1.1 Ink Disease

Phytophthora spp. first infect feeder roots that can be completely rotted away, and this results in an important loss of roots. In a further step of infection, larger roots are also colonized by the pathogen and show cortical necrosis or decay, with internal tissues turning brown or black. Roots of seedlings turn to be softer and easier to break than normal. For adult trees, root excavation can be very tricky, depending on the soil type and moisture content. Thus, most often only large roots and the tree collar or trunk base can be examined in adult trees. The necrosis of the cambial tissue caused by *Phytophthora* spp. may spread up to the tree collar and can be detected thanks to the black exudates which ooze out from bark cracks. When bark is removed, the necrosis appears as a brown and flame-shaped discoloration. Symptoms at the tree collar are generally sufficient to discriminate between ink disease and other root rots, including those caused by *Armillaria* spp. and *R. necatrix*. However, phenolic compounds oozing out can also be associated with wood borer insects and any bark injury.

11.2.1.2   Root Rots Caused by *Armillaria* spp. and *R. necatrix*

In the case of trees infected by *Armillaria* spp. and *R. necatrix*, necrosis of the cambial layer is less evident than in *Phytophthora* affected trees. Furthermore, either rhizomorphs and/or a thick white mycelium smelling of fresh mushroom or brown mycelial cords smelling of mold may be observed under the bark of main roots and basal parts of the trunk in the case of *Armillaria* spp. and *R. necatrix*, respectively. Typical *Armillaria* spp. fruiting bodies develop in autumn at the base or nearby infected trees.

Isolation of soil-borne pathogens (*Phytophthora* spp. or root rot fungi) can be done by plating small fragments of infected roots or bark tissues on selective nutritive media which allow the growth of the target pathogens while inhibiting that of most of the other microbes. Once *in vitro* cultures are obtained, molecular methods are used to identify the pathogen at the species level. For *Phytophthora* spp., another efficient technique involves baiting of propagules (spores) present in the soil with leaves, fruits, or seedlings of host plants. From these baits, isolation is then carried out for diagnostic purposes. On-site diagnostic tests are also available to detect and identify *Phytophthora* spp. in roots or bark collar tissues.

11.2.2   STEM AND BRANCHES

11.2.2.1   Chestnut Blight

Characteristic symptoms of chestnut blight are aerial cortical cankers. On smooth-bark stems and branches, cankers appear as red-orange necrotic areas, which eventually sink when the cambium is infected and the bark cracks. The older the tree is (and the thicker the bark is), the more inconspicuous the infection caused by *C. parasitica* is. Bark removal enables the observation of typical pale brown mycelial fans. Tree reaction results in a sunken and diffuse canker, in which host and fungal tissues are intimately intricate. With virus-free causal fungal strains, trees cannot heal the necrosis and cannot stop fungal spread in surface and in depth. Necrosis enlargement may lead to the girdling of the branch or stem and hence to the development of blight symptoms. Due to the alteration of the functional xylem in the infected organ, leaves of its distal part wilt and turn yellow and then brown, resulting in the typical flag symptom, i.e. sectorial death of the crown. At the same time, epicormic shoots develop at the base of the canker.

In coppices, high forests, and orchards, chestnut blight symptoms are observed at the base of the trunk, along the trunk, or on branches in the crown. On young grafted plants, infections are most frequently located in the region of the graft.

When the causal *C. parasitica* strains are virus-infected, i.e. the so-called hypo-virulent strains, tree defense reactions enable the healing of the canker. Mycelium spread is limited to the bark tissues, cambium, and xylem being protected by layers of bark and thus not colonized by the fungus. Such cankers have a swollen aspect. In addition, they are superficial and sometimes callused.

11.2.2.2   Other Shoot Blights and Cankers

The list of minor diseases of chestnut also include shoot blights and cankers associated either with *G. castaneae* or with *Sirococcus castaneae* (formerly known as *Diplodinia castanea*) that is the causal agent of the 'maladie de Javart', which was reported in France in the nineteenth century.

Bark cankers associated with *G. castaneae* look like those caused by *C. parasitica* but they can be differentiated by the color and aspect of fungal fruiting bodies and by the absence of typical mycelial fans of *C. parasitica* under the bark. They are frequent on young plants at the grafting point.

Symptoms caused by *S. castaneae* are deep, small, elongated cankers, starting most often from the tree collar. Infected areas are depressed and discolored. There is a risk of confusion with ink disease symptoms. Javart symptoms are reported most often on chestnut trees suffering from other stresses, either biotic or abiotic.

11.2.2.3   Wood Decays

Wood decays are commonly associated with mature and old trees, and they are caused by lignicolous basidiomycetes or, less frequently, ascomycetes. These fungi may be classified either as white rot or brown rot agents, depending on the component of the plant cell wall they are able to decompose, i.e. lignin or cellulose, respectively. Wood decay fungi reported on chestnut are listed in Table 11.1. The brown rot basidiomycetes *Daedalea quercina*, *Fistulina hepatica*, and *Laetiporus sulphureus* stand among the most widespread. Interestingly, *F. hepatica* has been recently reported in Spain associated with Chestnut Red Stain, a disease able to decrease the market value of chestnut timber to the point that coppices are no longer economical.

Wood decays are often poorly symptomatic. Diagnosis of wood decay fungi is traditionally based on the inspection of trees for the presence of fungal fruiting bodies and on their identification by using mycological keys. However, visual inspection is an inefficient diagnostic approach. Furthermore, fruiting bodies of some species are short-lived or may be found only in particular periods of the year.

TABLE 11.1

Tentative Frequency of the Main Wood Decay Fungi Reported on Chestnut

Fungal Species	Type of Rot	Frequency[a]
Antrodia albida	Brown	++
Antrodia serialis	Brown	+
Bjerkandera adusta	White	+++
Daedalea quercina	Brown	++++
Daedaleopsis confragosa	White	+++
Fistulina hepatica	Brown	++++
Fomitiporia robusta	White	+
Fomitopsis spraguei	Brown	+
Fuscoporia torulosa	White	++
Grifola frondosa	White	+++
Hapalopilus croceus	White	+
Junghuhnia nitida	White	++
Laetiporus sulphureus	Brown	++++
Lenzites betulina	White	+++
Meripilus giganteus	White	++
Perenniporia fraxinea	White	++
Polyporus corylinus	White	+
Polyporus umbellatus	White	++
Steccherinum oreophilum	White	−
Trametes hirsuta	White	+++
Trametes ochracea	White	++
Trametes versicolor	White	+++

[a] − very rare, + rare, ++ sporadic, +++ common, ++++ very common.

Therefore, the timing of diagnosis is also important. Most of wood decay fungi can be easily cultured from decayed wood, and therefore they could be identified through the use of appropriate keys, yet such an approach is difficult and time consuming. Early detection of some of the most important wood decay fungi of chestnut directly from wood samples is now possible by using DNA-based methods.

11.2.3 LEAVES

Powdery mildews produced by *Erysiphe* spp. become visible on the upper surface of leaves and on young shoots because of the appearance of a whitish-grey powdery looking coating which is formed of vegetative and sporulating hyphae. Besides leaf discoloration, heavy infections may cause growth distortion, premature leaf fall, and shoot drying.

Mycosphaerella maculiformis causes necrotic spots on both sides of chestnut leaves. At a first stage, in the early summer, they are initially pale, but then they become dark. They are of irregular shape and 1–2 mm in diameter but can coalesce and become larger in size. Depending on site and climatic conditions, the disease may be responsible of premature leaf shedding, leading to the weakening of the tree and to relevant reductions in the production of fruits.

Chestnut is reported in UK as a host plant for *Phytophthora ramorum* (https://www.forestresearch.gov.uk/tools-and-resources/pest-and-disease-resources/ramorum-disease-phytophthora-ramorum/). Chestnut leaves infected by *P. ramorum* show blackening and water soaking, with lesions originating from the leaf margins or tips and around spine-tipped teeth on the edge of leaves. There is little information on the distribution and impact of the disease.

Foliar symptoms of viral mosaic are reported on several chestnut varieties (Japanese, European, or hybrid varieties). They would be caused by a still unidentified virus called chestnut mosaic virus (ChMV[3]). ChMV results in chlorotic lesions on leaf veins expressed by yellow stripes and partial limb atrophy, but it can also induce stem cankers. The disease is mainly known as the driver of graft incompatibilities between an infected rootstock and a healthy scion.

11.2.4 FRUITS

In general, symptoms of nut rot are visible only once the fruit has been excised and the kernel exposed, although mummification caused by *Phomopsis* spp. may result in softer nuts compared to uninfected ones. Rots caused by *C. batschiana* may be recognized because the kernel becomes covered with a grayish mycelial mass. Later, both outer and inner kernel tissues become blackish and are characterized by an elastic texture. Conversely, when infected by *Phomopsis* spp., nuts turn chalky, sponge-like, and whitish.

Nuts colonized by *G. castaneae* exhibit discoloration and texture degradation typical of brown rots. A pale brown discoloration occurs at the margin of the endosperm, followed by loss of tissue consistency as it whitens and hardens, resulting in a chalky appearance. Few other fungi belonging to *Acremonium, Acrospeira, Alternaria, Amphiporthe, Cladosporium, Fusarium, Mucor, Penicillium,* and *Trichoderma* have been frequently isolatcd from harvested and infected nuts, which may create some confusion with the diagnosis of major nut rots. Rotted nuts usually float once they are immersed in water, and this may help discriminating them from unrotted and marketable ones.[4]

11.3 BIOLOGY AND EPIDEMIOLOGY OF PATHOGENS

11.3.1 INK DISEASE

Phytophthora spp. infect fine roots through zoospores (Figure 11.1).[5,6] These flagellate short-life asexual spores are released from sporangia (20–30 spores per sporangium). They are attracted by root exudates and can move, when free water is

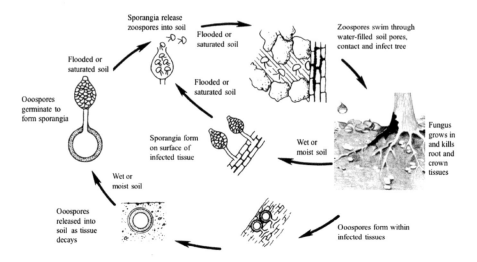

FIGURE 11.1   Disease cycle of soil-borne *Phytophthora* spp. responsible for the chestnut ink disease. (Courtesy of Cornell University Extension; Figure originally included in Wilcox, W.F., Phytophthora root and crown, Fruit Crops IPM Disease Identification Sheet No. 7. New York State IPM Program, 1992.)

present in the soil, towards the plant. At the contact point, they encyst, germinate, and penetrate in the root. This is the starting point of asexual cycles, which allow rapid (every 2 or 3 days under disease-conducive environmental conditions) and powerful amplification of the inoculum. As soon as roots are colonized by mycelia of *Phytophthora* spp., new sporangia are produced on the roots and new infections can occur if environmental conditions prevail. Sporangia formation has never been reported to occur at the tree collar and on trunk cortical tissues. *P. cambivora* and *P. cinnamomi* are heterothallic species, requiring the mating of two different mating types (or sexual types) to produce sexual progenies. However, sexual reproduction resulting in the development of oospores is hardly reported even in sites where the two mating types co-occur. *P. cinnamomi* can grow and persist in the soil or in infected plant material via chlamydospores, which are thick-walled asexual spores adapted for long term survival. *P. cinnamomi* has been reported to persist up to 6 years in moist soils. By contrast, chlamydospores have never been observed in the case of *P. cambivora*, which nonetheless can survive in nursery substrate soils up to 45 days.

Due to their life cycle, *P. cambivora* and *P. cinnamomi* are largely dependent on environmental conditions (especially rainfalls and temperature), cultural practices (irrigation, ground work), and site characteristics (topography and hydrography) for their natural dissemination, multiplication, survival, and infectivity. Human-assisted dissemination of both pathogens occurs through plant and soil movement in the frame of domestic and international commercial exchanges between plant nurseries or farms and of leisure activities (trekking and traveling). In the case of *P. cinnamomi*, long-distance dispersal proved to be very effective in relation to its broad host range, encompassing at least 3,000 ornamental and forest plant species.

11.3.2 CHESTNUT BLIGHT

Cryphonectria parasitica reproduces both asexually via conidiospores produced in pycnidia and sexually through ascospores produced in perithecia (Figure 11.2).[8] Both types of fruiting structures are embedded in stromas which develop within infected bark tissues. From these stromas several millions of conidiospores are produced during wet weather from the spring to the autumn and are dispersed by splashing and by rain over a short distance, i.e. within the same tree or to neighboring trees. Conidiopores can also be vectored via animals or tools at longer distances. In older stromas, ascospores are expelled in the air and are wind dispersed over several hundred of meters. Ascospores release occurs following rains. In France, the peak of ascospores was observed in May whereas it was reported in autumn in the United States. Both conidiospores and ascospores take advantage of fresh wounds, either natural or not, to penetrate and germinate within bark tissues. Susceptibility of *C. sativa* tissues to infections is highest during the growing

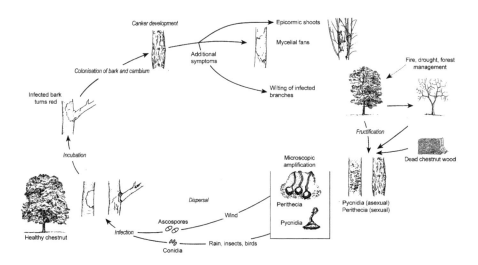

FIGURE 11.2 Life cycle of the chestnut blight pathogen *C. parasitica*.[9] (Courtesy of Simone Prospero and Daniel Rigling; Figure originally included in Prospero, S. and Rigling, D., Chestnut blight, In: *Infectious Forest Diseases*, CABI International, Wallingford, UK, pp. 318–339, 2013, adapted after the life cycle proposed by Ursula Heiniger.)

season, which also corresponds to the period of release of *C. parasitica* spores. The pathogen can stay inactive in the bark or spread in the cortical tissues as a latent or endophytic fungus.

11.3.3 Nut Rots

Some of the major nut rot agents of chestnut, including *C. batschiana*, *Phomopsis* spp., and *G. castaneae*, have been documented to live endophytically in some plant organs and tissues comprising, depending on the species, flowers, leaves, shoots, bark, and buds. Interestingly, these fungi may be found as endophytes in symptomless fruits still attached to the tree, suggesting that infection occurs on the tree rather than once the nuts have fallen.

The infection biology of *G. castaneae* has been almost elucidated.[2] Ascospores released from perithecia, harbored on rotten nuts and burrs and/or conidiospores released from acervuli formed on galls induced by the Asian gall wasp *Dryocosmus kuriphilus*, or on bark cankers, are responsible of infections occurring on flowers, leaves, and branches, although only those occurring on flowers will result in the development of nut rot. Based on the outcomes of a population genetics study conducted in Europe, the role of ascospores is prevailing compared to that of conidiospores in the infection biology of *G. castaneae*. Nut rot incidence at harvesting has been reported as directly significantly correlated to the levels of temperatures of the months preceding nut harvesting and models are available to estimate disease incidence.[2]

A similar infection biology, with airborne infections taking place at the flowering stage, has been hypothesized to arise for *C. batschiana*. However, for this and other nut rot fungi, post-harvest infection may either also occur or be the rule. Post-harvest infection may arise on the soil through damaged remnant of styles or wounds after the falling of nuts from trees in the orchard or during storage. All nut rot fungi are favored by high levels of relative humidity. However, *G. castaneae* and *Phomopsis* spp. are more thermophilic than *C. batschiana*, which may be frequent especially in cool climates.[1]

11.3.4 Other Diseases

While weakening may predispose trees to infection by some root rot fungi, including *A. bulbosa*, *A. mellea* and *R. necatrix* are regarded as primary pathogens attacking trees regardless of their level of vigor, the latter pathogen especially in soils characterized by high levels of moisture. *A. bulbosa* and *R. necatrix* may spread from colonized wood in the soil or from infected hosts to new ones through rhizomorphs and mycelial cords, respectively.[10] Besides, root contacts between infected and uninfected trees or contacts of roots with pieces of colonized wood in the soil are essential for infection and spread of *A. mellea*, which is characterized by fragile and short-lived rhizomorphs. Wounds are not crucial for root infection to occur, although they may aid in the case of *R. necatrix*.[10]

As for most foliar diseases, both *Erysiphe* spp. and *M. maculiformis* overwinter as mycelium or sexual fruiting bodies in fallen infected leaves, although the former can also overwinter as mycelium in twigs and buds.[10] Primary infections occur by means of wind-dispersed ascospores generally in the spring, while during the growing season the pathogen spreads mainly via conidiospores generated on infected tissues. Plants do not need to be stressed or injured for infections to occur. Climatic factors have an important role in disease expression. While dry periods or semi-arid environments may predispose plants to powdery mildews as conidiospores can germinate without extra moisture, leaf spots caused by *M. maculiformis* often develop in the autumn in association with a humid and cool climate. Hence, it is expected that this last disease may become severe especially at the bottom of cool valleys where air humidity accumulates.[10] Information on the biology and epidemiology of *P. ramorum* on chestnut are still limited, yet the pathogen has been shown to sporulate on leaves under laboratory conditions. Chestnut mosaic virus is transmitted either through grafting or by the aphid vector *Myzocallis castanicola*.[3]

In general, minor shoot blight and canker agents may gain access into the tree either directly through the stomata, leaf scars, and buds or through wounds, although an increasing body of literature points to several of them having endophytic behavior. The factors triggering the switch from the endophytic to the pathogenic stage and their relation to the onset of symptoms are mostly unknown, but they may be related to stressors, including climate change, resulting in the weakening of trees.

The biology of wood decay fungi in living trees has been previously reviewed.[11] In general, infections occur by means of airborne sexual spores through wounds, including pruning wounds or injuries on roots, tree collar, and stem. The larger the size of the wound, the higher the probability of successful infections by wood decay fungi, and this is also related to the fact that large size wounds may take years to occlude. Besides the above general scheme that applies to the most widespread wood decay fungi of chestnut, there is an increasing body of literature indicating that latent phases in symptomless tissues, hence endophytic phases, are also possible.

11.4   CONTROL

Control of chestnut diseases can be achieved by adopting one of the strategies described in the summary table reported in Technical Sheet 11.1, or by integrating them into Integrated Pest Management (IPM) systems.

TECHNICAL SHEET 11.1   MANAGEMENT STRATEGIES/TACTICS TO CONTROL CHESTNUT DISEASES

Management Strategies/Tactics		Ink Disease	Chestnut Blight	Nut Rots	Armillaria/ Rosellinia Root Rots	Mycosphaerella maculiformis Leaf Spots	Other Foliar Diseases	Other Cankers	Wood Decays
Avoidance	Avoid establishing plantations in infested and risky sites	✓				✓			
	Avoid injuries	✓		✓ᵃ				✓	✓
	Disinfect pruning/grafting tools		✓					✓?	✓
	Minimize size of pruning/grafting wounds		✓					✓?	✓ᵇ
	Timing of pruning		✓						✓
	Promptly collect fallen nuts and keep them dry in post-harvest			✓✓					
Exclusion	Clean vehicle (tires) or regulate their access to uninfested sites	✓							
	Prevent water flowing from infested to uninfested sites	✓✓							
	Disinfect equipmentᶜ	✓✓							
	Avoid using diseased and bad quality plants	✓✓	✓✓		✓✓				
	Quarantine regulations		✓✓						
Eradication	Remove/burn infected plants	✓✓ᵈ	✓✓	✓✓	✓✓ᵈ				
	Remove/bury/burn fallen burrs and pruning discards		✓	✓✓		✓		✓	✓
	Remove/bury/burn fallen leaves			✓✓		✓	✓		
	Remove/bury damaged nuts fallen onto the soil			✓✓					
Resistance		✓✓	(✓)	(✓)					

(Continued)

Disease Management Strategies/Tactics		Ink Disease	Chestnut Blight	Nut Rots	Armillaria/ Rosellinia Root Rots	Mycosphaerella maculiformis Leaf Spots	Other Foliar Diseases	Other Cankers	Wood Decays
Protection	Chemical protection	✓				✓	✓		
	Fertilization		✓			✓			
	Prophylactic pruning					✓			
	Application of wound dressing/seals		✓?						
Therapy	Biological control		✓✓						
	Therapeutic pruning		✓						
	Chemotherapy				✓				
	Hydrotherapy			✓					

✓✓ = highly effective and/or widely adopted; ✓ = effective and/or adopted; (✓) = under investigation but promising

[a] to nuts in post-harvest
[b] only pruning wounds
[c] in nurseries
[d] including root systems
? likely, but no supporting data available.

11.4.1 AVOIDANCE

This control strategy is aimed at avoiding environments or conditions that are conducive to the diseases. Locating plantations in areas that are unfavorable or marginal for pathogen growth and transmission may minimize disease. Hence, when establishing new orchards and plantations, to minimize the risk of occurrence of ink disease, sites characterized by well drained, loamy soils with high organic contents should be preferred. In fact, good drainage has been suggested to reduce the time that roots are in contact with free water, which is a required condition for pathogen infection, whereas high organic content has been reported to shorten the survival time of oospores and chlamydospores.[12]

To minimize the risk of occurrence of leaf spots, plantations should preferably not be established at the bottom of cool valleys where air humidity is likely to accumulate and to favor *M. maculiformis* infection. By contrast, soil type is not a good predictor for chestnut blight disease severity.

Pruning wounds may be easily infected by *C. parasitica* and wood decay fungi. These infections can be minimized by using disinfected tools, and by doing small wounds at the good time (i.e. in winter, when there is no aerial inoculum). Graft cuts are particularly vulnerable to *C. parasitica* and possibly to *G. castaneae*. Bud grafting may be less favorable to *C. parasitica* infection than bark grafting because it produces smaller wounds.[9]

To prevent infection by nut rot fungi in post-harvest, nuts fallen from trees onto the soil should be promptly collected by paying attention to avoid wounding or damaging them. If post-harvest treatments like immersion in cold or hot water are carried out, nuts should be immediately and carefully dried post-treatment to prevent fungal infections during storage.[4]

11.4.2 EXCLUSION

Exclusion encompasses quarantine regulations and any best management practice that may prevent pathogen introduction. Exclusion of ink disease may be achieved by applying the following procedures:

- Knowing where the pathogens are, using susceptible indicator species and soil baiting;
- Restricting vehicle movement from infested to uninfested areas;
- Cleaning vehicles before entering uninfested areas;
- Preventing water draining from infested to uninfested areas.

In particular, the establishment and a correct maintenance of canalization networks may help avoiding water flow on the soil and hence the spread of ink disease. In nurseries, avoid using contaminated irrigation water, and avoiding splashing and overwatering is recommended. In addition, the careful cleaning of all equipment may help in the exclusion of ink disease pathogens.

Diseases may also be prevented by using uninfected and vigorous plants in a new planting, because these conditions may not only prevent the introduction of new diseases, but also provide a greater chance for the survival of the nursery stock both to planting stress and to pathogens already present in the site, either primary or opportunistic parasites. Therefore, accurate inspections of plants should be conducted before planting by taking into account that, for some diseases, infected plants may be symptomless and that chemical treatments with systemic fungicides carried out in the nursery may have resulted in the remission of symptoms rather than in the pathogen eradication from plants.

Quarantine regulations are implemented across the world to prevent the spread and the introduction of *C. parasitica* into areas that are still free from the disease or of novel genotypes of the pathogen into areas already infested. Regulations regarding *C. parasitica* vary depending on the continent and country, but they generally affect the trade and movement of wood, bark, seeds, and living plants, and they require efficient diagnostics and surveillance measures. In Europe, regulations are in place against *C. parasitica* based on the Council Directive 2000/29/EC. Movement of living plants is regulated for chestnut and for oak, meaning that plants intended for planting of these plant genera can be moved within Europe only if they are accompanied by a plant passport.

11.4.3 ERADICATION

Eradication aims to remove a pathogen after its introduction but before it becomes widespread. However, since plant pathogens can rarely be fully eliminated, eradication is often intended as practices aimed at reducing pathogen inoculum and sanitation. The control of ink disease through eradication is particularly difficult to achieve because *Phytophthora* spp. can survive for several years as resting propagules in the soil. The removal of infected trees is thus not sufficient to eliminate the soil-borne inoculum. By contrast, in the case of root rot pathogens, including *Armillaria* spp. and *R. necatrix*, the removal of dead and diseased trees, comprising their stumps and root systems, may also have some effects in reducing pathogen inoculum.

Eradication of chestnut blight from orchards may be attempted to delay the epidemics and protect trees for a while. It is especially recommended if orchards are well apart from infested forest sites. Uprooting and burning of infected trees may be effective, although it may take decades for a successful eradication to be achieved.

The physical removal or the burial of the fallen burrs has been proposed as an effective control strategy for *G. castaneae* and *M. maculiformis* which infect and fruit on burrs. Pruning discards, especially those deriving from therapeutic pruning, should not be left on the soil to minimize the risk of spread of *C. parasitica* and other pathogens of stem and branches. For the same reason, all nuts should be promptly removed after harvest, especially those damaged and commonly left on the soil, to reduce the inoculum potential of nut rot agents like *G. castaneae*.[2,4]

11.4.4 PROTECTION

The protection aims to prevent pathogen infection by means of strategies as diverse as using biological, chemical, or spatial barriers, and/or by means of agronomic or silvicultural operations to prevent disease from occurring or to slow down the rate of spread of established diseases.

11.4.4.1   Chemical Protection

Both phenylamides, including the most common metalaxyl, and phosphonates, comprising fosetyl-aluminium and potassium salts of phosphorous acid, are systemic fungicides effective against *Phytophthora* spp.[13] The phosphonate potassium phosphite has proved to be effective in protecting trees from infections of *P. cinnamomi* when applied both as foliar sprays and stem injections. Even though foliar treatments were slightly less effective than injections, they may be more appropriate as preventative treatments in nurseries. While the repeated use of phenylamides may trigger the selection of pathogen strains resistant to the fungicides, pathogen resistance to phosphonates is deemed unlikely or slow to develop because of its complex mode of action, involving, depending on the dose, either a direct effect on the pathogen or an indirect effect on it through the stimulation of host defenses. To reduce the risk of selection of resistant pathogen strains, the combination with or alternation of metalaxyl with other fungicides is recommended.[12] Although copper sulfate as soil treatment has been reported as effective in reducing the infection rates of *P. cinnamomi*, copper compounds display a strictly protective function, having no effects once infection has occurred.[12]

Protection of chestnut from foliar diseases may be required especially in nurseries and may be achieved by spraying leaves with fungicides. Against powdery mildews, spraying with sulphur, triazole fungicides, organophosphates, and morpholines at intervals of two to three weeks when symptoms first occur in spring until August has proved to be effective. Leaf spots caused by *M. maculiformis* may be controlled by using either copper-related or systemic compounds.[10]

11.4.4.2   Fertilization

The fertilization of soils with nitrogen, phosphorus, and potassium has been reported to reduce the incidence of chestnut blight and the size of lesions on artificially inoculated Japanese chestnut trees in orchards.[9] In particular, high levels of nitrogen in the soil may promote the expression of resistance to blight. Mineral fertilization is also recommended against *M. maculiformis* leaf spots, especially where the disease affects valuable orchards.[10]

11.4.4.3   Protection Using Spatial Barriers

Protection from tree diseases may sometimes be achieved by optimizing the plantation density. A lack of association between the plantation density and the spatial pattern of nut rot caused by *G. castaneae* has been documented, suggesting that any attempt of controlling this pathogen by fine-tuning the orchard plantation density is likely to fail.[2]

11.4.4.4   Pruning

Intensive prophylactic pruning is recommended wherever *M. maculiformis* leaf spots display high severity, as this operation is expected to favor aeration through the crowns and hence to decrease the levels of relative humidity necessary for infections to occur.[10]

11.4.4.5   Other Protection Practices

A practice to protect trees from wound pathogens involves the application of treatments directly on the wound (e.g. graft, pruning cut) aimed at preventing infections and/or promoting a rapid sealing of the wound itself. However, wound dressing is generally ineffective in preventing infections by wood decay fungi, and no sound data exists so far about the efficacy of these treatments to prevent *C. parasitica* infections.[9]

11.4.5 RESISTANCE

Using Eurojapanese rootstocks is the most effective method for growning chestnuts in areas where *P. cambivora* and *P. cinnamomi* are widespread.[14]

Hybrid rootstocks or fruit varieties can also exhibit some resistance to *C. parasitica* in the field. However, there is no breeding program to improve chestnut blight resistance in Europe. By contrast, in the United States, the objective of two programs carried out by the American Chestnut Foundation (TACF) and the American Chestnut Cooperators Foundation (ACCF), initiated in the 1980s, was to restore the American chestnut *C. dentata* by interspecific hybridization with Chinese chestnut *C. mollissima*, backcrossing and genetic engineering.

Screening and testing host varieties for their resistance to nut rots are just beginning. In Australia, differences in the severity of symptoms caused by *G. castaneae* were detected among chestnut varieties depending on the biogeographical origin of the fungal strains used for the pathogenicity tests, despite the fact that all varieties were susceptible to the disease. In Europe, preliminary results suggest that the susceptibility profiles to nut rot caused by *G. castaneae* are comparable between the *C. sativa* wildtype and some chestnut varieties of local or global relevance.[2]

11.4.6 THERAPY

With therapy, diseased trees are cured once infection has occurred. In the case of chestnut, trees may be cured from some diseases by using chemicals (i.e. chemotherapy), therapeutic pruning, and possibly fertilizers, but the most interesting approach is known as the biological control of blight through hypovirulence.[8]

The rationale of this biological control method is to transmit CHV1 to the local *C. parasitica* strains through the establishment (i.e. inoculation) of virus-infected *C. parasitica* strains in cankers and hence to speed up with this human intervention the natural spread of hypovirulence observed in Italy since the 1950s. In order to maximize the success of the healing, virus-infected isolates should be chosen among those belonging to the same vegetative compatibility (vc) types of the local populations. Thus, before any field application, population studies must have been carried out and diversity of vc types assessed.

If the biological control through the inoculation of hypovirulent strains is not applied, therapeutic pruning aimed at removing cankers caused by virus-free *C. parasitica* strains should be carried out. However, healing cankers caused by hypovirulent strains should not be cut to maintain on site a local source of CHV1.

Potassium phosphite applied by means of trunk injections was reported effective as a therapeutic treatment against *P. cambivora* on adult chestnut trees: two applications were sufficient to promote plant recovery.[13] A combination of potassium phosphite and micronutrients solution injected into plants provided good therapeutic effects against *P. cinnamomi* as well.[15] However, treatment efficacy and the ability of trees to absorb the compounds were affected by the level of disease severity; the higher the level of disease severity, the lower the efficacy of treatment.[13] It should be noted that the level of absorption of a treatment into the tree when applied through injection/infusion depends upon the density and functionality of the canopy, the functionality of the vascular system, and all the climatic factors which can influence the tree transpiration rate. Hence, injections/infusions should be preferably carried out in sunny and possibly windy days to ensure that absorption occurs correctly.

Soil fertilization at the beginning of the tree growing season with manure and organic fertilizers has been suggested to show some therapeutic effects in trees displaying early symptoms of ink disease.[16]

Post-harvest treatments based on hydrotherapy consisting in the immersion of nuts in cold water for 5–8 days, by creating anaerobic conditions and subsequent fermentations, are expected to minimize the risk of development of moulds and nut rots.[4] Immersion of nuts in hot water combined with a treatment with cell-wall degrading enzymes deriving from a *Trichoderma harzianum* strain have proved promising in reducing losses caused by *G. castaneae*.[2]

11.4.7 INTEGRATED PEST MANAGEMENT AND REMARKS

Disease management practices are commonly more effective when applied as a system rather than individually, hence in the frame of IPM. Chestnut is threatened by several diseases and pests, and the choice of the most appropriate IPM option is largely dependent on their prevalence, although it may also be affected by technical constraints. Individual practices may be effective against some diseases and pests while being detrimental against others, and this should be taken into account. For instance, while the removal of fallen burrs and pruning discards from the soil is recommended to reduce infection rates of *G. castaneae*, and hopefully that of *M. maculiformis*, the same practice may be unbeneficial for the efficacy of the biological control against the Asian gall wasp *D. kuriphilus*, as it normally eliminates important reservoirs of the parasitoid *Torymus sinensis*. Therefore, decisions on the opportunity to apply certain practices should be taken case by case and require a broad knowledge encompassing not only plant pathology, but also entomology, horticulture, or silviculture.

In conclusion, a couple of comments and recommendations are necessary concerning phytosanitary products and especially chemical treatments. First, in order to be freely usable, a phytosanitary product, either chemical or biological, must be registered and approved for this use by the regulatory authorities operating at the country level. Second, as stated previously, chemical treatments carried out in nurseries may result in a remission of symptoms of the disease rather than in the eradication of the pathogen from plants. This implies a risk similar to that represented by infected plants remaining asymptomatic because of the endophytic behavior of pathogens. While both these conditions complicate the detection of the diseases, they are both expected to contribute to the spreading of chestnut pathogens.

REFERENCES

1. Conedera, M., Jermini, M. and Sassella, A. 2005. Raccolta, trattamento e conservazione delle castagne. Caratteristiche del frutto e principali agenti infestanti. *Sherwood* 107: 5–12.
2. Lione, G., Danti, R., Fernandez-Conradi, P., Ferreira-Cardoso, J., Lefort, F., Marques, G., Meyer, J., Prospero, S., Radócz, L., Robin, C., Turchetti, T., Vettraino, A.M. and Gonthier, P. 2019. The emerging pathogen of chestnut *Gnomoniopsis castaneae*: The challenge posed by a versatile fungus. *European Journal of Plant Pathology* 153: 671–685.
3. Desvignes, J. 1999. Sweet chestnut incompatibility and mosaics caused by the chestnut mosaic virus (ChMV). *Acta Horticulturae* 494: 451–458.
4. Conedera, M., Jermini, M., and Sassella, A. 2005. Raccolta, trattamento e conservazione delle castagne. Tecniche applicative e loro efficacia. *Sherwood* 107: 17–23.
5. Vannini, A. and Vettraino, A. 2011. *Phytophthora cambivora*. *Forest Phytophthoras* 1(1). doi:10.5399/osu/fp.1.1.1811
6. Robin, C., Smith, I. and Hansen, E. 2012. *Phytophthora cinnamomi*. *Forest Phytophthoras* 2(1). doi:10.5399/osu/fp.2.1.3041
7. Wilcox, W.F. 1992. Phytophthora root and crown rots. Fruit Crops IPM Disease Identification Sheet No. 7. New York State IPM Program.
8. Heiniger, U. 1994. Le chancre de l'écorce du châtaignier (*Cryphonectria parasitica*). Institut fédéral de recherches sur la forêt, la neige et le paysage.
9. Prospero, S. and Rigling, D. 2013. Chestnut blight. In: Gonthier, P. and Nicolotti, G., *Infectious Forest Diseases*, pp. 318–339. CABI International, Wallingford, UK.
10. Goidànich, G. 1990. *Manuale di Patologia Vegetale*. Edagricole, Bologna.
11. Vasaitis, R. 2013. Heart rots, sap rots and canker rots. In: Gonthier, P. and Nicolotti, G., *Infectious Forest Diseases*, pp. 197–229. CABI International, Wallingford, UK.
12. Hayden, K., Hardy, G.S.J., and Garbelotto, M. 2013. Oomycete diseases. In: Gonthier, P. and Nicolotti, G., *Infectious Forest Diseases*. CABI International, Wallingford, UK.
13. Gentile, S., Valentino, D. and Tamietti, G. 2009. Control of ink disease by trunk injection of potassium phosphite. *Journal of Plant Pathology* 91: 565–571.
14. Barreneche, T., Botta, R. and Robin, C. 2019. Advances in breeding of chestnuts. In: Serdar, Ü. and Fulbright, D., *Achieving Sustainable Cultivation of Tree Nuts*, Burleigh Dodds Science Publishing, in press.
15. Dal Maso, E., Cocking, J. and Montecchio, L. 2017. An enhanced trunk injection formulation of potassium phosphite against chestnut ink disease. *Arboricultural Journal* 39: 125–141.
16. Turchetti, T., Maresi, G., Nitti, D., Guidotti, A. and Miccinesi, G. 2003. Il mal dell'inchiostro nel Mugello (Fi): danni ed approcci di difesa. *Monti e Boschi* 1: 22–26.

12 Pests

Chiara Ferracini

CONTENTS

12.1 MAJOR AND MINOR PESTS, SYMPTOMS, CONTROL, AND BIO CONTROL

Many authors report chestnut pests as a variable number of 40 to 50 species, including mites and insects. These pests carry out their trophic activity mainly at the expense of the epigeal part such as trunk, branches, leaves, and fruits. Attacks to the roots of young plants in nurseries or new plants are occasionally reported, and mainly ascribed to the European mole cricket *Gryllotalpa gryllotalpa* L. (Orthoptera,

Gryllotalpidae) and the cockchafers *Melolontha hippocastani* F. and *M. melolontha* L. (Coleoptera, Scarabaeidae) (Figures 12.1 and 12.2).

Most of the species do not represent a serious threat and rarely account for a significant environmental and economic impact, therefore being considered occasional or minor pests. Conversely, the native pests developing at the expense of the chestnut fruit and the Asian chestnut gall wasp show greater potential to cause harm to chestnuts, and therefore they will be dealt with in a specific section in greater detail.

FIGURE 12.1 *Gryllotalpa gryllotalpa* L. adult. (Courtesy of DISAFA Entomology, University of Turin.)

FIGURE 12.2 *Melolontha melolontha* L. adult. (Courtesy of DISAFA Entomology, University of Turin.)

12.2 MAJOR PESTS

12.2.1 *Dryocosmus kuriphilus* Yasumatsu (Hymenoptera, Cynipidae)

One of the major key insect pests in chestnut orchards and forests is surely represented by *Dryocosmus kuriphilus* Yasumatsu (Hymenoptera, Cynipidae). This pest, commonly known as the Asian chestnut gall wasp (ACGW) or Oriental chestnut gall wasp (OCGW), is native to China and seriously affects all species belonging to *Castanea* genus (Fagaceae), namely the Japanese chestnut *C. crenata* Siebold & Zucc., the American chestnut *C. dentata* (Marsh) Borkh., the Chinese chestnut *C. mollissima* Blume and the European *C. sativa* Miller, and their hybrids. Moreover, two other gall wasp species, *Dryocosmus zhuili* n. sp. Liu et Zhu and *Synergus castaneus* PujadeVillar, Bernardo et Viggiani n. sp. (Hymenoptera, Cynipidae) have been recently described, inducing galls on *C. henryi* and *Castanea* spp., respectively.[1,2]

    D. kuriphilus involves serious damage on chestnut trees. This exotic species was established as a pest in the mid twentieth century in several countries, being reported in Japan, Korea, the United States, Nepal, and Canada. In Europe, *D. kuriphilus* is considered as one recent example of invasive alien species (IAS) accidentally introduced. It was first reported in Italy at the beginning of the twenty-first century and has rapidly spread throughout Europe (Figure 12.3).[3]

12.2.1.1 Description and Life Cycle

Adults are 3 mm long with black body and yellowish legs; they emerge in summer, from June to August, and survive 2 to 10 days, during which they search for chestnut buds to oviposit (Figure 12.4). This species has one generation per year, reproducing by thelytokous parthenogenesis. Each female lays on average 100 eggs. Eggs are oval, milky white, 0.1–0.2 mm long, with a long stalk. In each bud, a female lays 3–5 eggs, and several females may lay eggs into the same bud, making galls to reach up to 30 eggs. During the following vegetative season, the larvae

FIGURE 12.3 Distribution map of *Dryocosmus kuriphilus* Yasumatsu. (Courtesy of CAB International.)

FIGURE 12.4 *Dryocosmus kuriphilus* Yasumatsu female laying eggs inside a bud on sweet chestnut. (Courtesy of Phytosanitary Service, Regione Piemonte.)

develop within cells located in the inner tissues of the galls inducing the formation of greenish-red galls (see Chapter 13). Larvae are milky white, and pupae black or dark brown; both are 2.5 mm long when mature. *D. kuriphilus* most closely resembles the European oak cynipid wasp, *D. cerriphilus Giraud*, known to induce galls only on *Quercus cerris* L.[3]

12.2.1.2 Damage

Galls induced by *D. kuriphilus* develop in spring at the time of budburst in order to feed larvae during their development, and once mature present variable dimension, from 0.5 to 4 cm in diameter, in relation to the number of larval chambers. They are initially green with reddish shades, whereas after insect emergence galls desiccate and become woody, remaining on the tree for several years. Galls develop on shoots, leaves, and catkins, preventing shoot normal development, reducing leaf photosynthetic activity, and inhibiting flowering. A massive presence of galls results in a dramatic reduction of the chestnut vigour, causing significant yield losses. Indeed, this is one of the major pests attacking *Castanea* trees worldwide, responsible for a severe reduction in fruiting and negative impacts on chestnut production, with yield losses as high as 80%.[4] Severe infestations may also indirectly affect the plants, making trees more susceptible to other biotic factors, such as fungal infections. An increase in *Cryphonectria parasitica* (Murril) Barr, also known as chestnut blight, has been reported, and a severe epidemic of nut rot by

Gnomoniopsis castaneae Tamietti has spread in Italy and is currently regarded as a major pathogen of chestnut in France and Switzerland as well (for further details see Chapter 13).

12.2.1.3 Management Strategies

Although initially some authors reported destruction of galls before adult emergence, and systemic insecticides as control strategies, they proved ineffective mainly due to the huge presence of the gall wasp, to its fast dispersal, and to the location of chestnut trees in the forestry environment.

Initially in Japan, some varieties among the chestnut cultivars were found to be apparently resistant to the wasp in the field. The resistance mechanism was studied highlighting how in resistant varieties young larvae died in the larval chambers. Several new resistant varieties were bred and prevailed throughout Japan, and infestation by the cynipid was avoided for several years. However, a novel virulent strain of *D. kuriphilus* overcame plant resistance, and it spread widely.

Even in Europe, resistance was investigated. Most part of European chestnut varieties have shown to be susceptible to the pest, and a higher susceptibility has been observed for the hybrid 'Marsol' (*C. crenata* × *C. sativa*). On the contrary, in the Eurojapanese hybrid 'Bouche de Betizac', (*C. sativa* × *C. crenata*), plant cell necrosis occurred as soon as the eggs were laid. Even if larvae and eggs of the insect are found in the buds of this cultivar at the end of winter, no gall development occurs after budburst, supporting the hypothesis of the occurrence of a hypersensitive reaction as response to the cynipid infestation.[5] Furthermore, ACGW females were recently found to lay a significantly lower number of eggs in a *C. sativa* ecotype in southern Italy, highlighting fewer numbers of eggs laid, lower number of galls and larvae reaching the complete development, and smaller galls thus more susceptible to parasitism.[6]

Specific surveys have been carried out to investigate the control by indigenous parasitoids both in the area of origin and in the introduced areas. The community of native parasitoids recorded on invading ACGW populations is mainly composed of chalcid species (Hymenoptera, Chalcidoidea), commonly known to be parasitoids of oak cynipid gall wasps. Galls of *D. kuriphilus* develop midway between spring generation and the summer generation of native oak gall wasps, and thus when native parasitoids emerge, the developmental stage of chestnut galls may limit parasitism.

In its native area, its population is regulated by a complex of hymenopteran chalcid parasitoid species belonging to five different families, namely Eulophidae, Eupelmidae, Eurytomidae, Ormyridae, and Torymidae. In Japan the complex of parasitoids is much richer and presented by 24 species of hymenopteran chalcid parasitoid species belonging to seven different families (Cynipencyrtidae, Eulophidae, Eupelmidae, Eurytomidae, Ormyridae, Pteromalidae, and Torymidae). In Korea, after about 20 years from the introduction of the cinipid, 17 different species have been recruited.

The studies on native ACGW parasitoids available for Japan and the United States stated a rate of parasitism ranging from <1% to 8.5%, and around 1%, respectively.[7,8]

In Europe, instead, a more abundant literature reports parasitism rates up to 13% for Croatia, Slovenia, and Spain. In Italy, the community of native parasitoids recruited was evaluated in several regions, finding parasitism values quite similar to those reported for other European countries.[9,10] A remarkable parasitism rate was observed in 2011 in Northern Italy (Emilia Romagna region), mainly ascribed to *Torymus flavipes* (Walker) (Hymenoptera, Torymidae), though parasitism suddenly dropped in the following years.[11] In particular, the community associated with the exotic ACGW involved species shared with local populations of oak and rose gall wasps. Most of the recorded parasitoid species are commonly associated with the main oak galls (namely *Biorhiza pallida* Olivier, *Andricus quercustozae* Bosc, *A. lignicolus* Hartig, *A. curvator* Hartig, and *A. lucidus* Hartig).[9] However, none of these native species effectively controls the ACGW population in the long term, most likely due to incompatible life cycles.

Since chemical control, use of resistant chestnut varieties, and parasitism by natives were not feasible, the biological control approach was pursued. In particular, classical biological control (CBC) involves the selection of natural enemies of the invasive pest in their native range and releasing them in the invaded environment, aiming at long-term pest control. This control strategy has been extensively used, and as of 2006 there have been more than 7,000 introductions involving more than 2,600 invertebrate biocontrol agents, predominantly predatory, parasitic, and herbivorous insect, but also mites, nematodes, snails, and pathogens.[12]

With regard to the ACGW, the introduction of the most promising hymenopteran parasitoid, native to China, was performed. This exotic wasp is *Torymus sinensis* Kamijo (Hymenoptera, Torymidae), which was used for biological control initially in Japan, and then in the United States and Europe.

In Japan the parasitoid was imported from China, and release tests started in 1982. The parasitoid increased rapidly, and the gall density decreased to a level of only 3% infested buds by six years after the release.[13] In other sites a greater timing was needed, probably due to high mortality of the parasitoid associated with the activity of native facultative hyperparasitoids.

Since 1977, *T. sinensis* was released in the United States as well, proving to be able to establish and spread in the new area and becoming the most important species emerging from chestnut galls.[8]

With regard to Europe, the introduction of this BCA is considered as one of the most successful examples of recent European CBC. In Italy, *T. sinensis* was imported from Japan and released in 2005 in chestnut-growing areas as part of a biocontrol program funded by the Piedmont region.[14] In 2012, the Italian Ministry of Agricultural, Food and Forestry Policies (MiPAAF) actively pursued the national release of the parasitoid due to the evident impact on the decline in ACGW population. Following the positive Italian experience, further release programs were performed in Croatia, France, Greece, Hungary, Portugal, Slovenia, Spain, and Turkey.[15]

T. sinensis is reported in the literature as univoltine like its host, even if a prolonged diapause, mainly as late instar larva, has been proved, showing a two-year cycle.[16] It predominantly reproduces amphigonically. Adults are metallic green and 1.7–2.7 mm long (Figures 12.5 and 12.6).

1 mm

FIGURE 12.5 *Torymus sinensis* Kamijo adult female. (Courtesy of DISAFA Entomology, University of Turin.)

1 mm

FIGURE 12.6 *Torymus sinensis* Kamijo adult male. (Courtesy of DISAFA Entomology, University of Turin.)

FIGURE 12.7 *Torymus sinensis* Kamijo developmental stages: (a) larva; (b) pupa; (c) mature pupa (d) and adult female laying eggs inside a gall. (Courtesy of DISAFA Entomology, University of Turin.)

Females lay eggs into newly formed galls, usually one egg per host larva. After hatching, the larva feeds externally on the mature host larva until pupation, which occurs during late winter. Adult wasps emerge from withered galls in the spring, when chestnut trees start to sprout and with the appearance of *D. kuriphilus* galls (Figure 12.7).

T. sinensis is mass reared in controlled conditions (Figure 12.8). A detailed protocol was drawn up with the Italian Ministry of Agricultural, Food and Forestry Policies, with the aim to provide all the useful information for the release of the parasitoid, with regard to the collection of galls, mass rearing, and release in open field (Figures 12.8 and 12.9).

T. sinensis is a synovigenic species. Adults naturally feed, as all parasitoid wasps, on various sources in the field, such as floral nectar, homopteran honeydew, and pollen. Diet quality greatly influences survival and reproductive output, and wasps may reabsorb mature eggs to redirect energy towards other physiological needs such as survival, host seeking, and new egg maturation.

Younger females (2–3 weeks old) have to be preferred during the release stage in order to maximize parasitism.[17] Moreover, adults have the longest life span and significantly increase fecundity if honey plus pollen is provided as food, compared to pollen alone, water, and unfed wasps.[18]

This BCA is phenologically well synchronised with its host, and in all cases after its release it was able to disperse successfully alongside *D. kuriphilus* by expanding its

(a)   (b)

FIGURE 12.8   *Torymus sinensis* Kamijo specimens mass reared in controlled conditions (a) and released in infested chestnut growing area (b). (Courtesy of DISAFA Entomology, University of Turin.)

FIGURE 12.9   The biocontrol agent *Torymus sinensis* Kamijo released on highly infested sweet chestnut. (Courtesy of DISAFA Entomology, University of Turin.)

populations, reducing shoot infestation rates below the tolerable damage threshold, and significantly containing gall wasp outbreaks (Figure 12.10). The exponential growth reported in a 5–7-year period both for Japan and Italy make this parasitoid one of the most recent successful examples of classical biological control programmes.[13,15,19,20]

As opposed to Southwest Japan, no native European parasitoids negatively influenced the establishment of *T. sinensis*. Furthermore, *T. sinensis* so effectively controlled *D. kuriphilus* in Italy that its introduction progressively reduced the number of native parasitoids recruited since the establishment of the ACGW.[9]

(a)   (b)

FIGURE 12.10   Comparison between an infested and healthy branch. Pictures were taken
in the same chestnut growing area in NW Italy before the release of *Torymus sinensis* Kamijo
in 2004 (a) and after 8 years from the release of the biocontrol agent in 2013 (b). (Courtesy of
DISAFA Entomology, University of Turin.)

A major criticism of classical biocontrol is the lack of post-release impact evaluation
measures. Since *T. sinensis* is not native to Europe, several studies have investigated its
life history and reproductive traits,[16–18] assessing its host range,[21,22] its dispersal capabil-
ity,[20] and systematics deepening its relationship with closer and congeneric species.[23–25]

In Japan, evidence of hybridisation was reported between the introduced BCA
T. sinensis and the native *T. beneficus*.[23] Conversely, in the literature no mating was
ever recorded both in the field and in controlled conditions for Europe.[22]

Furthermore, the host-range of *T. sinensis* has been recently investigated.
In response to concern about non-target impacts associated with the introduction
of this exotic biological control agent, the EFSA Panel on Plant Health established
a new alternative host species list for testing the host-specificity of *T. sinensis*,
comprising galls which may be more susceptible to attack during the period that
females are searching for hosts (Figure 12.11). The species listed are *A. curvator*
sexual generation, *A. cydoniae* sexual generation, *A. grossulariae* sexual generation,
A. inflator sexual generation, *A. lucidus* sexual generation, *A. multiplicatus* sexual
generation, *B. pallida* sexual generation, *D. cerriphilus* sexual and asexual generation, *Neuroterus quercusbaccarum* (L.) sexual generation.

No choice trials were carried out applying protocols for host/prey range testing,
considering the hypothesis that species most closely related taxonomically and eco-
logically to the target are more likely to be utilized as hosts by the BCA being tested
(Figure 12.12). Furthermore, olfactometer bioassays were performed to assess the
attractiveness of volatiles for *T. sinensis* females.[21]

Field evidence for movement of *T. sinensis* to native oak galls was reported, and
17 non-target galls were found to be parasitised [*A. aries* (Giraud), *A. caputmedu-
sae* Hartig, *A. curvator, A. cydoniae, A. dentimitratus* Rejtö, *A. inflator, A. kollari*
Hartig, *A. lignicolus, A. lucidus, A. polycerus, A. quercustozae, Biorhiza pallida,
Cynips quercusfolii* L., *N. anthracinus* Curtis, *N. quercusbaccarum*, and *Synophrus
politus* Hartig].[22,25] However, at the present *T. sinensis* did not generate sufficient
mortality to imply some kind of population-level effect, and only *A. curvator* and

FIGURE 12.11   Some of the non-target gall species listed by EFSA for host-specificity testing after the release of the exotic parasitoid Torymus sinensis Kamijo: Biorhiza pallida Olivier (a), Andricus curvator Hartig (b), Andricus lucidus Hartig (c), Andricus grossulariae Giraud (d). (Courtesy of DISAFA Entomology, University of Turin.)

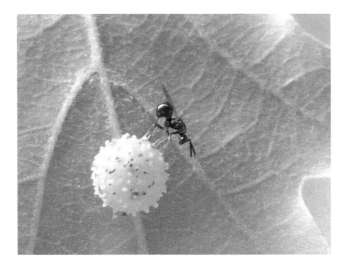

FIGURE 12.12   Oviposition of *Torymus sinensis* Kamijo in controlled conditions (no-choice test) on a non-target gall by *Andricus curvator* Hartig. (Courtesy of DISAFA Entomology, University of Turin.)

A. inflator proved to be more parasitised, suggesting a higher suitability for these non-target hosts.

Furthermore, a mathematical model describing the seasonal time evolution of the ACGW and *T. sinensis* has been developed.[26] According to this model, if both populations are able to produce an ever-changing pattern of travelling waves, minor and transitory risks of host-range expansion may occur on non-target hosts.[22]

12.2.2 LEPIDOPTERAN TORTRICID MOTHS (LEPIDOPTERA, TORTRICIDAE)

The European chestnut *C. sativa* is host for the chestnut leafroller (also called 'early chestnut moth') *Pammene fasciana* L., and the phylogenetically closely related beech tortrix *C. fagiglandana* (Zeller) (also called the 'intermediate chestnut moth'), and acorn moth or chestnut tortrix *Cydia splendana* (Hübner) (also called the 'late chestnut moth') (Lepidoptera: Tortricidae). They are considered as the key moth pests of chestnut in Europe, but they also share other common host plants. Indeed, north of the geographical distribution of chestnut, the larvae of *C. fagiglandana* feed on beech nuts (*Fagus sylvatica* L.) and the larvae of *C. splendana* on acorns (*Quercus* spp.). In southern Europe, both species are also found on chestnut, where they cause significant damage.[27] All these moths are oligophagous and monovoltine, with fruit-feeding (carpophagous) larvae developing in the fruit.

12.2.3 *PAMMENE FASCIANA* L.

12.2.3.1 Description and Life Cycle

P. fasciana adults have a wingspan of 14–17 mm long. Forewings are white with dark brown, blue-black, or blackish markings, and hindwings are greyish brown; they fly from early June onwards, with peak activity in late June and early July. Larvae, up to 13 mm long, cause an early fruit drop at the beginning of the development, but significant damage has rarely been reported. On the contrary, a serious threat is caused by the two other tortricid species (namely *C. fagiglandana* and *C. splendana*), whose flight activity takes place later in the season. These periods are more critical for chestnut because larvae with their trophic activity can damage the fruits directly. Moreover, a congeneric species *P. castanicola* Trematerra and Clausi, similar in morphology and biological features, was described from chestnut woods near Etna Vulcan (Sicily, southern Italy), and previously misidentified as *P. fasciana*.

12.2.4 *CYDIA FAGIGLANDANA* (ZELLER)

12.2.4.1 Description and Life Cycle

C. fagiglandana adult has a wingspan of 13–19 mm long. Forewing and hindwings are dark brown, with white markings. In the holm oak forests of southern Spain, adults emerge from late May to the end of October and egg laying occurs from late June to late October. In chestnut orchards, conversely to *P. fasciana*, *C. fagiglandana* adults are active later in the season from the end of July to the beginning of September. Usually one egg is laid inside the new shoot leaves near the acorns.

Larvae appear in acorns from middle July to late December and have an endophytic development inside the acorns lasting about 30–40 days. The mature larvae exit the acorns to overwinter in the ground or in bark cracks, protected by a cocoon (see also Chapter 13). Diapause period begins in late September or early October after the exit from acorns, and pupation occurs during May to September.

12.2.5 *Cydia splendana* (Hübner)

12.2.5.1 Description and Life Cycle

C. splendana is grey or dark-grey in colour, and the wingspan is 17–19 mm. The species overwinters in the last larval stage in a cocoon under the bark or in the soil until next spring. Adults appear in June–July in central-northern Europe, and in August in southern Europe.[28] Females lay up to 300 eggs usually on the leaves, and young larvae penetrate the fruit.

12.2.5.2 Damage

The larvae, with their trophic activity, cause premature drops of fruits, destruction of the cotyledons, and reduction in weight and size. They are responsible for great damage in chestnut production in Europe, being the cause for extensive economic losses annually. Fruit losses can range from up to 70% of harvested fruits depending on the year and plantation.

12.2.5.3 Management Strategies

In the literature, these pests have not been investigated to the same degree. *C. splendana* has received the most attention, followed by *C. fagiglandana*, with respect to aspects of biology and distribution and the identification of pheromone lures that could be used for monitoring.[29] *P. fasciana*, on the other hand, has always been considered to be a minor chestnut feeding pest.

Damage is variable in the different years and in the different sites. Chestnuts may be heavily attacked in some years and localities, while other times infestation is reported as scattered in chestnut orchards. In particular, due to the similarity in morphological traits and behaviour, these moths often cannot be reliably separated to species; that is why an accurate identification of pests is essential before applying any control strategy. Most procedures rely on morphological characterization of male genitalia and on molecular-based approaches. Nuclear ribosomal DNA (rDNA) and mitochondrial DNA (mtDNA) have been widely used for taxonomic and phylogenetic studies in insects, including tortricids, and thus may represent a rapid and reliable method to discriminate among species.

Since the chestnut leaf-roller is indeed a less frequent pest of chestnut compared with the other two *Cydia* species, control measures have been mostly targeted *C. fagiglandana* and *C. splendana*.

Chemical control is difficult because of the endophytic development of the larvae and the size of their tree hosts. Furthermore, protecting chestnuts with insecticide sprays is technically difficult due to the tree canopy, and few efficient insecticides are available.[30]

FIGURE 12.13   Pheromone-baited traps for monitoring the seasonal flight period of tortricid moths. (Courtesy of DISAFA Entomology, University of Turin.)

Control programs through pheromone applications seem to be the most important
control measure.

Pheromone-baited traps are increasingly important for monitoring the seasonal
flight period and population dynamics of these insects (Figure 12.13). Sex phero-
mone-related compounds of *C. fagiglandana* and *C. splendana* have been identified
from female gland extracts, (E, E)-8,10-dodecadien-1-yl acetate (E8E10-12:Ac) and
(E, E)-8,10-dodecadien-1-ol (E8E10-12:OH). In field experiments, blends of synthetic (E, E)- and (E,Z)-8,10-dodecadien-1-yl acetate were found to be attractive to
male moths,[31] and these two synthetic acetates as pheromone lures were successfully
applied for monitoring of the male chestnut tortrix.[32] Even so, variability in the
responses of *C. fagiglandana*, and *C. splendana* populations to pheromone blends
has been reported.[33] Trapping experiments revealed that traps baited with E8E10-
12:Ac are very attractive to males of *C. fugiglandana* and no but catch only a few
males of *C. splendana*, while traps with the alcohol (E, E)-8,10-dodecadien-1-ol
(E8E10-12:OH) catch low numbers of males of *C. fagiglandana* and no males of
C. splendana. (Z)-8-dodecen-1-yl acetate (Z8-12:Ac), which catches males of
Pammene fasciana (L.) and *P. fasciana* does not catch either of the two *Cydia*
species. When Z8-12:Ac is added to E8E10-12:Ac, the trapping power of the latter
substance for males of both *C. fagiglandana* and *C. splendana* becomes zero.[34]

Specifically, recent investigations with regard to the sex pheromone of
C. splendana from chestnut highlighted that *C. splendana* uses two different pheromone blends. As pointed out by Bengtsson et al.,[27] the female sex pheromone of
C. splendana feeding on acorns or oak nuts of *Quercus robur* in Sweden is a blend
of (E, Z)- and (E, E)-8,10-dodecadienylacetate (EZ and EE). However, the EZ/EE
pheromone blend of *C. splendana* from Swedish oak forests did not attract males in
chestnut plantations in France.

In Southern France, Hungary, and Switzerland, where *C. splendana* feeds on both
chestnut *C. sativa* and several species of oak, males respond instead to a blend of *E*,
E- and Z8, *E*10-12Ac (*EE/ZE*).

In addition, as application of these findings, a novel approach was recently proposed, which combines the basic principles of mating disruption with puffers, an innovative pheromone-dispensing method. Though encouraging, these data are just preliminary and therefore the 'puffer approach' needs further trials to be set up and optimised. Moreover, the application of pheromones in monitoring, mating disruption, and control of *Cydia* species is still under evaluation in northern Italy (Ferracini, unpublished data).

Several studies have been conducted to examine and optimize the use of control measures involving semiochemicals affecting moth behaviour. Volatile compound from ripe pear, ethyl (E, Z)-2,4-decadienoate (pear ester, PE), was described as a strong kairomonal attractant for both sexes of a closely related tortricid, the codling moth, *Cydia pomonella* L, and attraction of *C. splendana* to PE was also reported.[33] Furthermore, recent findings suggest that the addition of acetic acid (AA) proved to act synergistically on *C. pomonella* and some other moths belonging to the Tortricidae (*Hedya nubiferana* Hübner) and Sesiidae families (*Synanthedon myopaeformis* Borkhausen), increasing activity of PE for *C. splendana* as well. The response to PE in *C. splendana* could indicate a role in food-finding behaviour, or in the location of oviposition sites, even if it remains unknown whether pear ester is emitted by any part of the chestnut tree.

Traditional practices to manage burr and nuts in post-harvest are still a useful way to control these pests, and in the literature treatments such as water curing are reported. In particular, warm and cold baths were successful in eliminating *C. splendana* larvae, though the cold bath was inadequate to control *C. elephas*.[35]

To reduce the larval populations in the following season, chestnut fruits affected and fallen to the ground have to be collected and removed in order to prevent larvae entering the soil to overwinter. Furthermore, the use of entomopathogenic nematodes to be administered to the ground when population of cydia are in the wintering larval stage has been evaluated. One of the most successful approaches for soil insect pests is the use of entomopathogenic nematodes (EPN)s. In particular, *Steinernema feltiae* Filipjev, *S. weiseri* Mracek, Sturhan & Reid (Rhabditida, Steinernematidae), and *Heterorhabditis bacteriophora* Poinar (Rhabditida, Heterorhabditidae) have been tested against last instar larvae of *C. splendana*, and in soil pot experiments at 15°C, revealed *S. weiseri* as the most virulent species.[36]

12.2.6 *Curculio elephas* Gyllenhal (Coleoptera, Curculionidae)

12.2.6.1 Description and Life Cycle

With regard to chestnut weevil, *Curculio elephas* Gyllenhal (Coleoptera, Curculionidae) is reported as the main pest of the European chestnut. Moreover, further minor species, namely *C. propinquus* (Desbrochers) and *C. glandium* Marsham, have been reported developing on oak acorns and less frequently on chestnut.[30]

The chestnut weevil, *C. elephas*, is a univoltine species also developing in the fruit (acorns) of oak trees (*Quercus* spp.). It is generally distributed in the United States, Europe, and parts of North Africa and the Middle East.[28] Adult weevils are grey-yellow in colour and are 6–10.5 mm long, with a typical and distinctive snout on the head (Figure 12.14). They emerge from the soil in August and September,

FIGURE 12.14 *Dryocosmus kuriphilus* Yasumatsu female. (Courtesy of DISAFA Entomology, University of Turin.)

and the females lay eggs on or in the chestnuts where the larvae feed on the kernel for about 2 months. Conversely to tortrix moths, larvae are curved in the typical C-shape (see Chapter 13). In October, the last instar larvae, 10–15 mm long, emerge through a small hole made in the pericarp, drop to the ground, and enter the soil to a depth of 5–15 cm where they build a small earthen chamber within which they overwinter. Chestnut weevils can prolong diapause, enabling the pest to stagger adult emergence, with some adults emerging in the first summer after entering the soil while others emerge the following year or subsequent years.[37] However, as with the other life stages, this depends on the geographical region.

12.2.6.2 Damage

Young larvae dig a tunnel into the fruit and usually leave after harvesting, moving in the soil. An emergence hole may be easily detected on the fruits. Holes by *C. elephas* larvae are larger than those of *Cydia* spp. (average hole diameter: 1.2–1.5 mm for *Cydia* spp., and 3–4 mm for *Curculio* spp.).

12.2.6.3 Management Strategies

Even if insecticide treatments may allow a lower infestation rate at the harvest, the restriction on the use of many pesticides, particularly in protected areas, requires alternative products.

Use of nets under the canopy may reduce the potential infestation rate for the later years, and infested fruits need to be destroyed in storage. In the literature, BCAs such as entomopathogenic fungi (EPF) and EPNs are considered as an efficient and cost effective method of pest control, reducing emergent adults irrespective of the length of larval diapauses.[37]

Some preliminary tests on the ability of different strains and species of EPNs (namely *Heterorhabditis* and *Steinernema*) to infect these insect pests have been investigated, highlighting different results among EPN strains and the target species. Moreover, the EPF *Metarhizium brunneum*, used alone and in combination with the EPN *H. bacteriophora*, causes high mortality of both chestnut weevil and chestnut tortrix *C. splendana*. The efficacy of *S. feltiae*, *S. weiseri*, and *H. bacteriophora* on both species has been evaluated in Turkey in soil pot experiments.[36] *S. weiseri* revealed the most virulent species against *C. elephas* and *C. splendana*, with a higher susceptibility to the nematode for the latter, as previously described.

12.3 MINOR PESTS

12.3.1 APHIDS

Lachnus roboris L. and *Myzocallis castanicola* (Baker) (Hemiptera: Aphididae) are the two most important species for chestnut woods and plantations. *L. roboris*, also called black aphid, forms brown-blackish populations on younger branches, on suckers, and sometimes on green husks. *M. castanicola*, also called yellow aphid, is usually found in the lower part of the crown, at the bottom of the suckers. In case of high presence of both species, a general debilitation of the plant is recorded due to their feeding activity. Furthermore, the production of honeydew may favour the development of sooty mould fungi. Infestations are usually rare, and no control measure is adopted. In case of heavy infestations in nurseries and young plantations, mineral oils can be applied in presence of the overwintering eggs.

12.3.2 MOTHS

Over 10 lepidopteran pests are reported on chestnut as minor pests. Among the most common and widespread pests that in some years cause occasional defoliation, the European goat moth *Cossus cossus* L. (Lepidoptera, Cossidae) is noteworthy. Despite being known as a secondary pest, the goat moth is recorded occasionally in chestnut orchards, and its density has recently increased in Turkey. Damage was reported to be higher in orchards infected with the chestnut blight *C. parasitica*.[38]

   C. cossus is a destructive insect pest widely distributed throughout the Palaearctic region. It is a polyphagous species, feeding on a large number of plant species. Larvae inflict serious damage to broad-leaved trees, in particular poplar cultivations, and fruit trees by tunnelling into their trunks. Young larvae are light pinky reddish, and when mature they are reddish on the dorsal side and yellow on the ventral side (see Chapter 13).[28] Larvae overwinter in tunnels burrowed into trunks or branches. Adult wingspan is generally 70–100 mm, and forewings are grey with a creamy colour at the base and marked with wavy cross lines.[28] They fly from July onwards, and females lay about 700 eggs on the tree trunk. After egg eclosion, larvae penetrate through the bark, where they overwinter. Then mature larvae abandon the host plant and go into the soil, where they build a cocoon. Larvae are very destructive, feeding on the cambium, phloem, xylem, and parenchyma tissues. One generation is

FIGURE 12.15   The gypsy moth *Lymantria dispar* L. (Courtesy of DISAFA Entomology, University of Turin.)

completed every 2–3 years.[28] Even if reports are occasional, pyramidal funnel traps are usually used to monitor the presence of the pest, and biotechnological strategies by pheromones seems to be a sustainable way of control of the infestations.

Moreover, other frequently encountered species are the oak leaf roller moth (*Tortrix viridana* L.) (Lepidoptera, Tortricidae), the winter moth (*Operophtera brumata* L.) (Lepidoptera, Geometridae), and the buff-tip moth *Phalera bucephala* L. (Lepidoptera, Notodontidae) feeding on the parenchyma of several broad-leaved leaves, including chestnut. Even the gypsy moth *Lymantria dispar* L. (Lepidoptera, Lymantriidae), although developing preferably at the expense of oaks, can cause periodical defoliation of the chestnut trees as well (Figure 12.15).

12.3.3   AMBROSIA BEETLES

Ambrosia beetles (Coleoptera, Curculionidae Scolytinae) can be important pests of nursery production. The Asian *Xylosandrus germanus* (Blandford) and *X. crassiusculus* (Motschulsky) were the dominant species attacking chestnut in the United States in years 1998–2000,[39] and very intense attacks by *X. compactus* (Eichhoff) on young plants of *Castanea* sp. were observed in China in 1995–2000.[40] The congeneric species *X. germanus* was recorded on *C. crenata* in Japan as well. Also, the European shot-hole borer *Xyleborus dispar* F. has been reported in chestnut growing areas, especially in relation to the increased presence of stressed plants.

Furthermore, the granulate ambrosia beetle *X. crassiusculus* is considered as originating from tropical and subtropical Asia. The species usually performs two generations per year. Adults and larvae bore into twigs, branches, or small trunks of woody host plants and introduce a symbiotic ambrosia fungus (*Ambrosiella* sp.) on which they feed. Unlike other ambrosia beetles, which normally attack only stressed or damaged plants, *X. crassiusculus* is apparently able to attack healthy plants. Infested plants can show wilting, branch dieback, shoot breakage and general decline.

When boring galleries, frass is pushed out in the form of a compact cylinder which may reach 3 to 4 cm long (see Chapter 13). Adults are small dark reddish brown scolytids (female: 2–3 mm long, males: 1.5 mm). Larvae are white, legless, C-shaped with a well-developed capsule, and cannot be easily distinguished from other scolytids.

X. crassiusculus is a widely polyphagous species, able to colonize at least one hundred species belonging to various genera of forest, agricultural, and ornamental trees, including *Castanea* genus. It was reported for the first time in Europe infesting chestnut trees in NE Italy, and very recently in NW Italy, as developing in Japanese hybrids of chestnut trees growing in specialized cultivations.

In addition, large infestations of the black stem borer *X. germanus* were observed in various young plantations in NW Italy, probably due to an unexpected thermal increase recorded in spring.

The beetles are difficult to control. Mass trapping with funnel traps baited with ethanol may be used in chestnut orchards, and prompt clear-cut and destruction of the infested trees is needed to contain the damage and reduce the pest population density.

12.3.4 Mites

Many specimens of eriophyd mites have been collected from chestnut trees in Japan, namely *Aceria japonica* Huang, *Phaulacus acutilobus* and *P. obtusilobus* Keifer, and *Coptophylla matsudoensis* Keifer (Acari: Eriophyidae), but no noticeable damage has been observed.[41] In Japan and China, *Castanea* is considered as host plant for the spruce spider mite *Oligonychus ununguis* Jacobi (Acari: Tetranychidae) as well. In laboratory conditions, host acceptance of the sweet chestnut *C. sativa* by *O. ununguis* was tested, excluding the possibility for the pest to develop.[42]

Furthermore, specimens of four species of corticolous mites, *Liebstadia humerata* Sell. (Oribatida: Protobatidae), *Scheloribates latipes* (C. L. Koch) (Oribatida, Scheloribatidae), *Thyreophagus corticalis* (Mich.) (Acaridida, Acaridae), and *Zygoribatula laubieri meridionalis* Travé (Oribatida: Oribatulidae), were found in cankers, caused by *Cryphonectria parasitica* on chestnut trees, using mycelium of the fungus as food.[43]

In Europe two species have been specifically reported associated with chestnut: the tetranichid mites, *Oligonychus bicolor* (Banks) and occasionally *Eotetranychus carpini* Oudemans. The former was found for the first time in Europe (Sicily, southern Italy) on chestnut and oak. On chestnut leaves the mite lives along the midrib and lateral veins, causing a clear discolouration of the infested areas due to its feeding activities. Silken threads, secreted by the mites, are easily detectable on the host plants. Although the presence of the mites, no deformation nor premature leaf drop have been reported.

12.3.5 Accidental Introduction of Exotic Pests

Although most exotic introduced species are relatively innocuous, biological invasions by non-native species can cause irreversible economic and ecological impacts.

In particular, in consideration of the intense trade in goods with the United States, China, and other Eastern countries, accidental introduction and/or invasion by exotic insect pests to potentially suitable areas has been steadily increasing in recent years with considerable consequences for agriculture and forest ecosystems.

More than 150 exotic insect pests species feeding on *Castanea* spp. were listed in a recent review by Sabbatini Peverieri et al.[44] Sixty-two percent are native to East Asia, particularly China but also Japan and Korea, while the other 38% lives in North America (mainly the United States). The defoliating insects are the more consistent trophic group (38%), whereas the seed-feeders, the xylophagous, the sap-sucking, and all the others are in order less represented.

Among the exotic species at greater risk of introduction and able to develop on chestnut, only few species seem to actually represent a serious threat. First of all, the chestnut phylloxerid *Moritziella castaneivora* Miyazaki (Homoptera, Phylloxeridae) is considered an invasive species, introduced to China via seedling transportation from Japan or Korea. *M. castaneivora* infests flowers and fruits of *C. crenata* and *C. mollissima* and causes yellowed chestnuts, stunted growth, and premature fruit drop.[44] According to the information available in the literature, infested propagating material may act as the main pathway of introduction of the pest, thus representing an important threat to European chestnut cultivation.

Furthermore, the two-lined chestnut borer *Agrilus bilineatus* (Weber) (Coleoptera, Buprestidae) was one of the main pests causing the American chestnut (*C. dentata*) death in the United States, and its recent report in the Belgrad forest in Turkey increases the risk for sweet chestnut as well.

Similar to Europe, the genus *Curculio* includes in Asia and North America some of the most harmful chestnut-associated seed-feeding insects, such as *Curculio bimaculatus* Faust, *C. sikkimensis* (Heller), and *C. davidi* (Fairmaire) native to East Asia, together with *C. caryatrypes* (Boheman) and *C. say* (Gyllenhal) recorded in the United States. Moreover, the lepidoptera *Cydia glandicolana* (Danilevsky), and *C. kurokoi* (Amsel) are also reported to be noxious pests requiring adequate control strategies.[44]

Lastly, it is worth pointing out how some exotic insect pests recently introduced into Europe may represent a serious threat to broadleaved trees, and thus potentially to chestnut.

The citrus longhorn beetle (CLB), *Anoplophora chinensis* (Forster) (Figure 12.16) and the Asian longhorn beetle (ALB) *A. glabripennis* (Motshulsky) (Coleoptera, Cerambycidae) affect a wide range of broadleaved trees and shrubs. CLB may infest species in the family Fagaceae (beech, chestnut, oak). In particular, *C. crenata* is listed as species host for CLB, even if no specific information on the development of the pest is provided.

Both wood-boring beetles have been accidentally introduced into Europe, CLB in 2000 and ALB in 2007 respectively. Studies were carried out on maturation feeding and oviposition of CLB on the sweet chestnut *C. sativa* under quarantine conditions.[45] These preliminary results showed that *C. sativa* plants were accepted both for feeding and oviposition, but no complete development of the pest was ever recorded.

Also the brown marmorated stink bug, *Halyomorpha halys* (Stål) (Hemiptera, Pentatomidae), is an exotic pest recently introduced into Europe that deserves attention. The pest, native to Eastern Asia, is an extremely polyphagous species, causing

FIGURE 12.16   The citrus longhorn beetle, *Anoplophora chinensis* (Forster). (Courtesy of DISAFA Entomology, University of Turin.)

FIGURE 12.17   The brown marmorated stink bug, *Halyomorpha halys* (Stål). (Courtesy of DISAFA Entomology, University of Turin.)

significant damage both to agricultural crops and ornamental plants (Figure 12.17). It was recorded for the first time in Europe (Switzerland) in 2004, and it is now spreading rapidly worldwide. In recently invaded areas in Europe and Eurasia, including northern Italy and western Georgia, severe damage has been observed especially in pear and hazelnut orchards.

In Asia, *H. halys* has had a confusing nomenclatural record, having previously been referred to as *H. mista* or *H. brevis* (synonyms), and *H. picus* (misidentification), but currently all references to *Halyomorpha* spp. are considered synonymous with *H. halys*.[46] In the literature, *H. picus* was reported as a potential pest for chestnut trees in China.[44] Hence, even if detailed information is lacking, it may represent a serious threat to *Castanea* genus.

The rate at which humans translocate species beyond their native ranges has substantially increased during the last centuries, and the increase in numbers of alien species does not show any sign of saturation at a global scale.[47]

What determines the likelihood of herbivores to incorporate new hosts in their diet remains controversial.[48] Rapid evolution of host range has been documented in several species. In particular, exotic pests may adapt to non-target plants, especially during outbreaks, thus increasing damage to new host plants, with significant implications in the agroecosystems.

Given the high rate of introduction for exotic species, a constant monitoring is needed to evaluate the capacity of these species to exploit new hosts, expanding the ecological host range, and in order to apply the most effective control strategies in a timely manner.

REFERENCES

1. Zhu, D.-H., Liu, Z., Lu, P.-F., Yang, X.-H., Su, C.-Y. and Liu, P. 2015. New gall wasp species attacking chestnut trees: *Dryocosmus zhuili* n. sp. (Hymenoptera: Cynipidae) on *Castanea henryi* from Southeastern China. *Journal of Insect Science* 15(1): 156.
2. Bernardo, U., Gebiola, M., Xiao, Z., Zhu, C.-D., Pujade-Villar, J. and Viggiani, G. 2013. Description of *Synergus castaneus* n. sp. (Hymenoptera: Cynipidae: Synergini) associated with an unknown gall on *Castanea* spp. (Fagaceae) in China. *Annals of the Entomological Society of America* 106: 437–446.
3. CABI. Invasive species compendium. https://www.cabi.org/isc/datasheet/20005 (June 13, 2019).
4. Battisti, A., Benvegnù, I., Colombari, F. and Haack, R.A. 2014. Invasion by the chestnut gall wasp in Italy causes significant yield loss in *Castanea sativa* nut production. *Agricultural and Forest Entomology* 16(1): 75–79.
5. Sartor, C., Dini, F., Torello Marinoni, D., Mellano, M.G., Beccaro, G.L., Alma, A., Quacchia, A. and Botta, R. 2015. Impact of the Asian wasp *Dryocosmus kuriphilus* (Yasumatsu) on cultivated chestnut: Yield loss and cultivar susceptibility. *Scientia Horticulturae* 197: 454–460.
6. Nugnes, F., Gualtieri, L., Bonsignore, C.P., Parillo, R., Annarumma, R., Griffo, R. and Bernardo, U. 2018. Resistance of a local ecotype of *Castanea sativa* to *Dryocosmus kuriphilus* (Hymenoptera: Cynipidae) in Southern Italy. *Forests* 9: 94.
7. Murakami, Y. and Gyoutoku, Y. 1995. A delayed increase in the population of an imported parasitoid, *Torymus* (*Syntomaspis*) *sinensis* (Hymenoptera: Torymidae) in Kumamoto, southwestern Japan. *Applied Entomology and Zoology* 30: 215–224.
8. Cooper, W.R. and Rieske, L.K. 2011. A native and an introduced parasitoid utilize an exotic gall-maker host. *Biological Control* 56: 725–734.
9. Ferracini, C., Bertolino, S., Bernardo, U., Bonsignore, C.P., Faccoli, M., Ferrari, E., Lupi, D., Maini, S., Mazzon, L. and Nugnes, F. 2018. Do *Torymus sinensis* (Hymenoptera: Torymidae) and agroforestry system affect native parasitoids associated with the Asian chestnut gall wasp? *Biological Control* 121: 36–43.
10. Quacchia, A., Ferracini, C., Nicholls, J.A., Piazza, E., Saladini, M.A., Tota, F., Melika, G. and Alma, A. 2013. Chalcid parasitoid community associated with the invading pest *Dryocosmus kuriphilus* in North-Western Italy. *Insect Conservation and Diversity* 6: 114–123.

11. Panzavolta, T., Croci, F., Bracalini, M., Melika, G., Benedettelli, S., Tellini, F.G. and iberi, R., 8078049. 2018. Population dynamics of native parasitoids associated with the Asian Chestnut Gall Wasp (*Dryocosmus kuriphilus*) in Italy. *Psyche: A Journal of Entomology* Article ID 8078049.

12. Hajek, A.E., Hurley, B.P., Kenis, M., Garnas, J., Bush, S.J., Wingfield, M.J., van Lenteren, J. and Cock, M.J. 2016. Exotic ecological control agents: A solution or contribution to arthropod invasions? *Biological Invasions* 18: 953–969.

13. Moriya, S., Shiga, M. and Adachi, I. 2003. Classical biological control of the chestnut gall wasp in Japan, pp. 407–415. *Proceedings of the 1st International Symposium on Biological Control of Arthropods*. USDA Forest Service, Washington, DC.

14. Quacchia, A., Moriya, S., Bosio, G., Scapin, G. and Alma, A. 2008. Rearing, release and settlement prospect in Italy of *Torymus sinensis*, the biological control agent of the chestnut gall wasp *Dryocosmus kuriphilus*. *Biological Control* 53: 829–839.

15. Avtzis, D.M., Melika, G., Matošević, D. and Coyle, D. 2019. The Asian chestnut gall wasp *Dryocosmus kuriphilus*: A global invader and a successful case of classical biological control. *Journal of Pest Science* 1: 107–115.

16. Ferracini, C., Gonella, E., Ferrari, E., Saladini, M.A., Picciau, L., Tota, F., Pontini, M. and Alma, A. 2015a. Novel insight in the life cycle of *Torymus sinensis*, biocontrol agent of the chestnut gall wasp. *Biological Control* 60: 169–177.

17. Picciau, L., Ferracini, C. and Alma, A. 2017. Reproductive traits in *Torymus sinensis*, biocontrol agent of the Asian chestnut gall wasp: Implications for biological control success. *Bulletin of Insectology* 70: 49–55.

18. Picciau, L., Alma, A. and Ferracini, C. 2019. Effect of different feeding sources on lifespan and fecundity in the biocontrol agent *Torymus sinensis*. *Biological Control* 134: 45–52.

19. Ferracini, C., Ferrari, E., Pontini, M., Saladini, M.A. and Alma, A. 2019. Effectiveness of *Torymus sinensis*: A successful long-term control of the Asian chestnut gall wasp in Italy. *Journal of Pest Science* 9: 353–359.

20. Matošević, D., Lacković, N., Kos, K., Kriston, E., Melika, G., Rot, M. and Pernek, M. 2017. Success of classical biocontrol agent *Torymus sinensis* within its expanding range in Europe. *Journal of Applied Entomology* 141: 758–767.

21. Ferracini, C., Ferrari, E., Saladini, M.A., Pontini, M., Corradetti, M. and Alma, A. 2015b. Non-target host risk assessment for the parasitoid *Torymus sinensis* in Italy. *Biological Control* 60: 583–594.

22. Ferracini, C., Ferrari, E., Pontini, M., Hernández Nova, L.K., Saladini, M.A. and Alma, A. 2017. Post- release evaluation of the non-target effects of *Torymus sinensis*, a biological control agent of *Dryocosmus kuriphilus* in Italy. *Biological Control* 62: 445–456.

23. Yara, K., Sasawaki, T. and Kunimi, Y. 2010. Hybridization between introduced *Torymus sinensis* (Hymenoptera: Torymidae) and indigenous T. beneficus (late-spring strain), parasitoids of the Asian chestnut gall wasp *Dryocosmus kuriphilus* (Hymenoptera: Cynipidae). *Biological Control* 54: 14–18.

24. Montagna, M., Gonella, E., Pontini, M., Ferrari, E., Ferracini, C. and Alma, A. 2018. Molecular species delimitation of the Asian chestnut gall wasp biocontrol agent released in Italy. *Insect Systematics and Evolution* 50: 327–345.

25. Pogolotti, C., Cuesta-Porta, V., Pujade-Villar, J. and Ferracini, C. 2019. Seasonal flight activity and genetic relatedness of Torymus species in Italy. *Agricultural and Forest Entomology* 21: 159–167.

26. Paparella, F., Ferracini, C., Portaluri, A., Manzo, A. and Alma, A. 2016. Biological control of the chestnut gall wasp with *T. sinensis*: A mathematical model. *Ecological Modelling* 338: 17–36.

27. Bengtsson, M., Boutitie, A., Jósvai, J., Toth, M., Andreadis, S., Rauscher, S., Unelius, C.R. and Witzgall, P. 2014. Pheromone races of *Cydia splendana* (Lepidoptera, Tortricidae) overlap in host plant association and geographic distribution. *Frontiers in Ecology and Evolution* 2: 46.
28. Vacante, V. and Kreiter, S. 2014. *Handbook of Pest Management in Organic Farming.* CABI, Wallingford, UK, 504 p.
29. Avtzis, D.N. 2012. The distribution of *Pammene fasciana* L. (Lepidoptera: Tortricidae) in Greece: An underestimated chestnut-feeding pest. *International Journal of Pest Management* 58(2): 115–119.
30. Avtzis, D.N., Perlerou, C. and Diamandis, S. 2013. Geographic distribution of chestnut feeding insects in Greece. *Journal of Pest Science* 86: 185–191.
31. Witzgall, P., Chambon, J.-P., Bengtsson, M., Unelius, C.R., Appelgren, M., Makranczy, G., Muraleedharan, N. et al. 1996. Sex pheromones and attractants in the Eucosmini and Grapholitini (Lepidoptera, Tortricidae). *Chemoecology* 7: 13–23.
32. Jósvai, J.K., Voigt, E. and Tóth, M. 2016. A pear ester-based female-targeted synthetic lure for the chestnut tortrix, *Cydia splendana. Entomologia Experimentalis et Applicata* 159: 370–374.
33. Schmidt, S., Anfora, G., Ioratti, C., Germinara, G.S., Rotundo, G. and De Cristofaro, A. 2007. Biological activity of ethyl (E, Z)-2,4-decadienoate on different tortricid species: Electrophysiological responses and field tests. *Environmental Entomology* 36(5): 1025–1031.
34. Den Otter, C.J., De Cristofaro, A., Voskamp, K.E. and Rotundo, G. 1996. Electrophysiological and behavioural responses of chestnut moths, *Cydia fagiglandana* and *C. splendana* (Lep., Tortricidae), to sex attractants and odours of host plants. *Journal of Applied Entomology* 120: 413–421.
35. Jermini, M., Conedera, M., Sieber, T.N., Sassella, A., Schärer, H., Jelmini, G. and Höhn, E. 2006. Influence of fruit treatments on perishability during cold storage of sweet chestnuts. *Journal of the Science of Food and Agriculture* 86(6): 877–885.
36. Karagoz, M., Gulcu, B., Hazir, S. and Kaya, H.K. 2009. Laboratory evaluation of Turkish entomopathogenic nematodes for suppression of the chestnut pests, *Curculio elephas* (Coleoptera: Curculionidae) and *Cydia splendana* (Lepidoptera: Tortricidae). *Biocontrol Science and Technology* 19: 755–768.
37. Asan, C., Haziz, S., Cimen, H., Ulug, D., Taylor, J., Butt, T. and Karagoz, M. 2017. An innovative strategy for control of the chestnut weevil *Curculio elephas* (Coleoptera: Curculionidae) using Metarhizium brunneum. *Crop Protection* 102: 147–153.
38. Kaplan, C. and Turanlı, T. 2018. Seasonal fluctuations of goat moth (*Cossus cossus* L.) in chestnut orchards in İzmir and Manisa, Turkey. *Acta Horticulturae* 1220: 103–108.
39. Oliver, J.B. and Mannion, C.M. 2001. Ambrosia beetle (Coleoptera: Scolytidae) species attacking chestnut and captured in ethanol-baited traps in Middle Tennessee. *Environmental Entomology* 30(5): 909–918.
40. Yan, S.P., Huang, H.Y. and Wang, J.B. 2001. The occurrence of chestnut beetle and its control. *South China Fruits* 30(1): 48.
41. Kadono, F. 1988. Three new species of eriophyid mites from chestnut trees in Japan (Acarina: Eriophyidae). *Applied Entomology and Zoology* 23: 150–155.
42. Czajkowska, B. and Puchalska, E. 2006. European larch (Larix decidua) and sweet chestnut (*Castanea sativa*) as host plants of spruce spider mites (*Oligonychus ununguis* Jacobi). *Biological Letters* 42: 307–313.
43. Nannelli, R., Turchetti, T. and Maresi, G. 1998. Corticolous mites (Acari) as potential vectors of *Cryphonectria parasitica* (Murr.) barr hypovirulent strains. *International Journal of Acarology* 24: 237–244.
44. Sabbatini Peverieri, G., Binazzi, F. and Roversi, P.F. 2017. Chestnut-associated insects alien to Europe. *Journal of Zoology* 100: 103–113.

45. Sabbatini Peverieri, G. and Roversi, P.F. 2010. Feeding and oviposition of *Anoplophora chinensis* on ornamental and forest trees. *Phytoparasitica* 38: 421–428.

46. Lee, D.-H., Brent, D., Short, B.D., Shimat, V.J., Bergh, J.C. and Leskey, T.C. 2013. Review of the biology, ecology, and management of *Halyomorpha halys* (Hemiptera: Pentatomidae) in China, Japan, and the Republic of Korea. *Environmental Entomology* 42(4): 627–641.

47. Seebens, H., Blackburn, T.M., Dyer, E.E., Genovesi, P., Hulme, P.E., Jeschke, J.M., Pagad, S. et al. 2017. No saturation in the accumulation of alien species worldwide. *Nature Communications* 8: 14435.

48. Castagneyrol, B., Jacte, H., Brockerhoff, E.G., Perrette, N., Larter, M., Delzon, S. and Piou, D. 2016. Host range expansion is density dependent. *Oecologia* 182(3): 779–788.

13 Pests, Diseases, and Physiological Disorders *Vademecum*

*Alberto Alma, Gabriele Beccaro, Chiara Ferracini,
Paolo Gonthier, Cécile Robin, and Ümit Serdar*

CONTENTS

13.1 LEAVES

Chlorosis/leaves yellowing or scorched at the edge

(a)

Courtesy of Gamba, G.

(b)

Courtesy of DISAFA Plant Pathology, University of Turin.

Visual Assessment

Leaves produce insufficient chlorophyll. Chlorotic leaves are pale, yellow, or yellow-white (a). Severe deficiencies may result in leaves scorched at the edges and in leaf curl (b).

Causes and Remedies

Physiopathological disorders related to: specific mineral deficiencies in the soil or substrate, such as iron, magnesium, or zinc; too high pH in soil or substrate; poor drainage.

In the case of chestnuts in pot, the substrate is not adequate. A fertilization with iron, in the form of a chelate or sulphate, magnesium or nitrogen compounds in various combinations could be helpful to solve the deficiency.

Red tips

Courtesy of Gamba, G.

Visual Assessment	Causes and Remedies
The young apical leaves become reddish starting from their tips and in 10–20 days can become necrotic.	This physiological disorder can occur in different conditions: excessive salinity of the substrate (or irrigation water) or growing in greenhouse or tunnel with excessive heat or radiation.

Powdery mildews

Courtesy of DISAFA Plant
Pathology, University of Turin.

Courtesy of INRA, UMR BIOGECO.

Visual Assessment	Causes and Remedies
A whitish-gray powdery looking coating appears on the upper surface of leaves and on young shoots. Besides leaf discoloration, the disease may cause growth distortion, premature leaf fall, and shoot drying.	Powdery mildews are caused by the foliar fungal pathogens *Erysiphe* spp. In nurseries, spraying leaves with sulphur, triazole fungicides, organophosphates, and morpholines at intervals of two to three weeks when symptoms first occur in spring until August may be effective. Removal of fallen leaves from the soil before infections of new leaves occur in spring may have some effects.

Necrotic leaf spots

Courtesy of DISAFA Plant
Pathology, University of Turin.

Courtesy of INRA, UMR
BIOGECO.

Visual Assessment	Causes and Remedies
Necrotic spots initially pale, but then darker, of irregular shape, and 1–2 mm in diameter become visible on both sides of the leaves starting from the early summer. Spots may coalesce and become larger in size.	Necrotic spots are caused by the foliar fungal pathogen *Mycosphaerella maculiformis*. Removal of fallen leaves and burrs from the soil before infections of new leaves occur in spring may have some effects. In severely affected orchards, intensive prophylactic pruning is recommended as well as mineral fertilization. Foliar treatments with copper-related compounds may be effective in preventing infections.

Gall wasp

Courtesy of DISAFA Entomology,
University of Turin.

Courtesy of DISAFA Entomology,
University of Turin.

Courtesy of DISAFA Entomology,
University of Turin.

Visual Assessment

Greenish-red galls developing mainly on shoots and leaves, in spring at the time of budburst. At the end of summer galls desiccate and become woody, remaining on the tree for several years.

Causes and Remedies

Galls are induced by the Asian chestnut gall wasp *Dryocosmus kuriphilus* Yasumatsu. Adult females lay eggs in chestnut buds during summer. Only in the following vegetative season, the larvae develop inducing the formation of the galls.

The exotic pest can be controlled, releasing the biocontrol agent *Torymus sinensis* Kamijo. This exotic parasitoid must be released when chestnut trees sprout, in presence of *D. kuriphilus* greenish-red galls (usually April–May).

13.2 STEM AND BRANCHES

Sun burn

Visual Assessment	Causes and Remedies
The bark of young trees darkens in association with the necrosis of bark tissues. It usually occurs exclusively on only one side of the tree, the most exposed to the sun.	Damage can occur when solar irradiation is too high. Plastic shelters could cause irradiation concentration. Protect the trunk with white paint; don't use plastic shelters.

Courtesy of Fabro, M.

Graft incompatibility

Visual Assessment	Causes and Remedies
Regression in vegetative growth of the scion, yellowing shoots, symptoms of water stress, drying of the scion, overgrowth at the graft union.	Graft incompatibility

Courtesy of Serdar Ü.

Chestnut blight

(a)

Courtesy of DISAFA Plant
Pathology, University of Turin.

(b)

Courtesy of DISAFA Plant
Pathology, University of Turin.

Visual Assessment

On smooth-bark stems and branches, cankers appear as red-orange necrotic areas, which eventually sink when the cambium is infected and the bark cracks. Deep longitudinal cracks appear and epicormic shoots develop at the base of the canker (a). Enlargement may lead to the girdling of the branch or stem and hence to the development of blight symptoms.

Superficial cankers, sometimes callused and with a swollen aspect may also be observed (b). These last cankers are rarely associated with blight symptoms.

Causes and Remedies

Blight and cankers are caused by the fungal pathogen *Cryphonectria parasitica.* Superficial and healing cankers are associated with *C. parasitica* strains bearing a mycovirus, i.e. hypovirulent strains.

Therapeutic biological control may be achieved by inoculating cankers with hypovirulent *C. parasitica* strains, although this strategy is effective only if information on the pathogen's local population structure is available.

Control strategies also include:

- Pruning branches showing shoot blight, deep bark cracks, and epicormics shoots while keeping superficial and callused cankers
- Minimization of the size of grafting wounds and application of wound seals
- Disinfection of pruning/ grafting tools
- Prevention of injuries on trees from happening during the host growing season
- Removal of pruning discards
- Fertilization
- The use of healthy and high quality nursery plants

Wood decays

(a)

Courtesy of DISAFA Plant Pathology, University of Turin.

(b)

Courtesy of DISAFA Plant Pathology, University of Turin.

Visual Assessment

Wood decays often develop into the heartwood and hence they are poorly symptomatic, unless they appear in association with wounds or if fruiting bodies of decay agents are present (a, *Fistulina hepatica*). Affected wood either shrinks, shows a brown discoloration, and cracks into roughly cubical pieces (b, brown rot) or becomes moist, soft, spongy, or stringy, its color turning to whitish (white rot).

Causes and Remedies

Wood decays are caused by lignicolous basidiomycetes or, less frequently, ascomycetes, infecting trees mostly by means of airborne spores through wounds. They can remain viable as saprophytes and fruit on dead wood.

Disease control is preventative and includes the following measures:

- Minimization of injuries to the trees
- Minimization of the size of pruning wounds
- Disinfection of pruning tools
- Removal of dead wood and pruning discards

Xylosandrus spp. and *Xyleborus*

Courtesy of Dutto, M.

Visual Assessment

Infested plants can show wilting, branch dieback, shoot breakage, and general decline. When larvae bore the galleries, frass is pushed out in the form of a compact cylinder which may reach 3–4 cm long.

Causes and Remedies

Damage can be caused by *Xylosandrus germanus* (Blandford), *X. crassiusculus* (Motschulsky), *X. compactus* (Eichhoff), and *Xyleborus dispar* (F.).

The beetles are difficult to control. Mass trapping with funnel traps baited with ethanol may be used in chestnut orchards, and prompt clearcut and destruction of the infested trees is needed to contain the damage and reduce the pest population density.

Reddish granular mass, general decline

Courtesy of Gamba, G.

Visual Assessment

Infested plants can show wilting, branch dieback, and general decline. Presence of reddish granular mass formed of sawdust and frass mixed with silk can be visible at the foot of the tree.

Causes and Remedies

Damage is caused by the European goat moth *Cossus cossus* L. Pheromone traps are used to monitor the population in chestnut orchards. Control measures are difficult since larvae are protected inside the trunk. Entomopathogenic nematodes have been tested both in laboratory and field conditions, but results have not been promising, and research is still ongoing.

Aphids

Courtesy of Beccaro G.

Visual Assessment

Brown-blackish or yellow aphids, depending on the species. Populations usually affect younger branches, suckers, and sometimes the green husks.

Causes and Remedies

Damage is caused by the black aphid *Lachnus roboris* L. and the yellow aphid *Myzocallis castanicola* (Baker).

They are considered as occasional pests, but infestations may occur especially in nurseries and young plantations. No control measure is usually adopted. In case of heavy infestations mineral oils can be applied in presence of the overwintering eggs.

13.3　ROOTS AND TREE COLLAR

Ink disease

(a)

Courtesy of INRA, UMR
BIOGECO.

(b)

Courtesy of INRA, UMR
BIOGECO.

(c)

Courtesy of Deplaude, H., CA
Ardèche, France.

Visual Assessment

First symptoms are root infections, which can be easily observed on seedlings. Infected roots show cortical necrosis or decay, with internal tissues turning brown or black (a).

Consequences of root infections are tree dieback, which starts with chlorosis and defoliation, and mortality (b).

In some trees, the necrosis of the cambial layer spreads up to the tree collar and trunk base, becoming apparent when bark is removed (c). Black exudates can ooze out from bark cracks at the trunk base.

Causes and Remedies

Ink disease is caused by the fungi-like organisms *Phytophthora cambivora* and *P. cinnamomi.*

Disease control is mostly preventative, although some therapeutic measures are also available, and includes the following:

- Check of the quality of nursery stock
- Use of resistant rootstocks
- Establishment of plantations in well drained, loamy soils with high organic contents
- Minimization of injuries to the roots and tree collar
- Prevention of water flowing from infested to uninfested sites through the establishment and correct maintenance of canalization networks
- Wash of vehicles before accessing to uninfested sites
- Removal of infected/dead trees, including their root systems, although care should be taken to avoid the spread of inoculum
- Copper sulphate as a soil treatment but also phosphonates as foliar sprays or stem injections have proved effective in reducing pathogen infection rates. Trunk injections with potassium phosphite also display some therapeutic effects.

Armillaria *and* Rosellinia *root rots*

Courtesy of DISAFA Plant
Pathology, University of Turin.

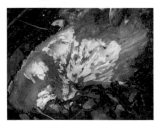

Courtesy of DISAFA Plant
Pathology, University of Turin.

Visual Assessment

Either rhizomorphs (see figure for *Armillaria rhizomorphs*) and/or a thick white mycelium smelling of fresh mushroom (see figure for *Armillaria mycelium*) or brown mycelial cords smelling of mould may be observed under the bark of main roots and basal parts of the trunk in the case of *Armillaria* spp. and *Rosellinia* spp., respectively. Typical *Armillaria* spp. fruiting bodies may also be observed in autumn at the base or nearby infected trees.

These diseases are associated with wood decay of roots and often with tree dieback and mortality.

Causes and Remedies

These fungal diseases are caused either by *Armillaria* spp. or by *Rosellinia necatrix*. Disease control is preventative and includes the following measures:

- Check of the quality of nursery stocks
- Removal of infected/dead trees, including their root systems

13.4 NUTS

Chestnut moth and weevil

Courtesy of DISAFA Entomology, University of Turin.

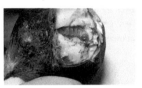

Courtesy of DISAFA Entomology, University of Turin.

Courtesy of DISAFA Entomology, University of Turin.

Courtesy of DISAFA Entomology, University of Turin.

Courtesy of DISAFA Entomology, University of Turin.

Visual Assessment

Presence of larvae developing in the fruit, causing premature drops, destruction of the cotyledons, and a reduction in weight and size. A depression in the hilum area may occur, highlighting the presence of galleries caused by larval internal feeding. Larvae dig galleries that are filled with fine excrement. When mature larvae exit the fruit, a typical round hole is visible on the pericarp (diameter 1.5–4.0 mm, depending on the species).

Causes and Remedies

Damage can be caused by four species of spermophagous pests: the chestnut leafroller (also called 'early chestnut moth') *Pammene fasciana* L., the beech tortrix *C. fagiglandana* (Zeller) (also called the 'intermediate chestnut moth'), the chestnut tortrix *Cydia splendana* (Hübner) (also called the 'late chestnut moth'), and the chestnut weevil *Curculio elephas* Gyllenhal.

Adult flight activity is checked by pheromone lures.

The tortricid moths *Cydia* spp. are commonly controlled by mating disruption. Collection of infested chestnut fruits fallen to the ground can prevent larvae entering the soil to overwinter. Moreover, techniques such as cold water soaking, hot water soaking, steam treatment, and low oxygen storage can be used as well. Entomopathogenic nematodes (*Steinernema* spp. and *Heterorhabditis bacteriophora*) and fungi (*Beauveria bassiana*, *Metarhizium anisopliae*) can be administered to the ground when population are in the wintering stage.

Nut rots

(a)

Courtesy of DISAFA Plant Pathology,
University of Turin.

(b)
Courtesy of DISAFA Plant Pathology,
University of Turin.

(c)
Courtesy of DISAFA Plant Pathology,
University of Turin.

Visual Assessment

Symptoms of nut rot are generally visible only once the fruit has been excised and they vary depending on the disease agent. Rots caused by *Ciboria batschiana* result in kernels becoming blackish and characterized by an elastic texture (a), while rots caused by *Phomopsis* spp. in nuts turning chalky, sponge-like, and whitish (b). Nut rot caused by *Gnomoniopsis castaneae* result in a pale brown discoloration at the margin of the endosperm, followed by loss of tissue consistency as it whitens and hardens, resulting in a chalky appearance (c).

White hilum

Courtesy of Beccaro G.

Visual Assessment

The hilum is white and wet. The nuts quickly dehydrate in the post-harvest.

Causes and Remedies

Nuts are not completely ripe. Maybe burs were collected directly from the plant or fell too early.

Index

Note: Page numbers in italic and bold refer to figures and tables, respectively.

T - #0817 - 101024 - C380 - 234/156/17 - PB - 9781032084305 - Gloss Lamination